22nd International Congress on X-Ray Optics and Microanalysis 2013

Journal of Physics: Conference Series Volume 499

Hamburg, Germany
2 - 6 September 2013

ISBN: 978-1-63266-724-3
ISSN: 1742-6588

Printed from e-media with permission by:

Curran Associates, Inc.
57 Morehouse Lane
Red Hook, NY 12571

Some format issues inherent in the e-media version may also appear in this print version.

Printed by Curran Associates, Inc. (2014)

For permission requests, please contact the Institute of Physics
at the address below.

Institute of Physics
Dirac House, Temple Back
Bristol BS1 6BE UK

Phone: 44 1 17 929 7481
Fax: 44 1 17 920 0979

techtracking@iop.org

Additional copies of this publication are available from:

Curran Associates, Inc.
57 Morehouse Lane
Red Hook, NY 12571 USA
Phone: 845-758-0400
Fax: 845-758-2634
Email: curran@proceedings.com
Web: www.proceedings.com

TABLE OF CONTENTS

INSTRUMENTS AND TECHNIQUES

APPLICATIONS

Journal of Physics: Conference Series **499** (2014) 011001 doi:10.1088/1742-6596/499/1/011001

Preface

The 22nd edition of the International Congress on X-ray Optics and Microanalysis (ICXOM 22) was held from 2–6 September 2013, in Hamburg, Germany. The congress was organized by scientists from DESY in collaboration with TU Dresden and Helmholtz-Zentrum Geesthacht, who also formed the scientific advisory board. The congress was hosted in the historical lecture hall building of the University of Hamburg located in the city center.

ICXOM22 was attended by about 210 registered participants, including 67 students, and was open for listeners. The attendance was split between 26 countries (Germany 120, rest of Europe 57, America 20, Asia 8, Australia 6).

The ICXOM series is a forum for the discussion of new developments in instrumentation, methods and applications in the fields of micro- and nano-analysis by means of X-ray beams. Following the trend of the last 10 years, the conference focusses more and more on synchrotron radiation rather than X-ray laboratory sources. Besides micro-beam X-ray fluorescence and absorption spectroscopy, different methods based on diffraction and full-field imaging were covered. Newly introduced to the ICXOM series was scanning coherent X-ray diffraction imaging, which was shown to evolve into a mature method for the imaging of nanostructures, defects and strain fields. New developments on fast X-ray detectors were discussed (*Lambda, Maia*) and advances in X-ray optics — like the generation of a sub 5nm point focus by Multilayer Zone plates — were presented. Talks on micro- and nano-analysis applications were distributed in special sessions on bio-imaging, Earth and environmental sciences, and Cultural heritage.

The congress featured nine keynote and ten plenary talks, 56 talks in 14 parallel sessions and about 120 posters in three afternoon sessions. Seventeen commercial exhibitors exposed related X-ray instrumentation products, and two luncheon seminars on detector electronics were given. This allowed us to keep the student fees low and to distribute eight student travel grants. The Wednesday was devoted to an outing to DESY with guided tours to PETRA III and FLASH experiments and to the European XFEL construction site. A lecture was given by Henry Chapman introducing to structural imaging at X-ray free-electron lasers. Talks highlighting the current status and future of nanoanalysis at the leading synchrotron facilities APS (J Maser), ESRF (P Cloetens) and SPRing8 (Ishikawa) were given in the DESY auditorium offering other DESY scientists the opportunity to follow the talks.

Figure 1. Participants of ICXOM22 gathered in front of the main building of the University of Hamburg.

Further information about ICXOM22, including a detailed program and electronic abstract book, can be found on the congress website www.icxom22.de.

We thank all the participants of ICXOM22, everybody who helped in the organization and are looking forward to hearing about further progress during ICXOM23, which will be organized by Brookhaven National Laboratory in Uptown, New York.

Gerald Falkenberg
ICXOM22 conference chair

Christian Schroer
ICXOM22 co-chair

Peer review statement

All papers published in this volume of *Journal of Physics: Conference Series* have been peer reviewed through processes administered by the proceedings Editors. Reviews were conducted by expert referees to the professional and scientific standards expected of a proceedings journal published by IOP Publishing.

Journal of Physics: Conference Series **499** (2014) 012001 doi:10.1088/1742-6596/499/1/012001

Maia X-ray Microprobe Detector Array System

D P Siddons[1], R Kirkham[5], C G Ryan[4], G De Geronimo[1], A Dragone[2], A J Kuczewski[1], Z Y Li[3], G A Carini[2], D Pinelli[1], R Beuttenmuller[1], D Elliott[1], M Pfeffer[1], T A Tyson[3], G F Moorhead[5], P A Dunn[5]

[1] Brookhaven National Laboratory, Upton, NY 11973, USA.

[2] SLAC National Accelerator Laboratory, 2575 Sand Hill Road, Menlo Park, CA 94025, USA

[3] New Jersey Institute of Technology, Newark, NJ 07102, USA

[4] CSIRO Earth Sciences and Resource Engineering, 26 Dick Perry Avenue, Kensington, WA 6151, Australia

[5]CSIRO Materials Science and Engineering, Normanby Road, Clayton VIC 3168, Australia

siddons@bnl.gov

Abstract. Maia is an advanced system designed specifically for scanning x-ray fluorescence microprobe applications. It consists of a large array of photodiode detectors and associated signal processing, closely coupled to an FPGA-based control and analysis system. In this paper we will describe the architecture and construction of the system.

1. Introduction

As x-ray sources become more powerful, the fluorescence detector is increasingly the system bottleneck in x-ray microprobe systems. Maia addresses this by providing a very large array of small detectors and adopting a novel data acquisition and analysis strategy. Most present microprobe systems either do 'step-and-acquire' measurements, acquiring an energy spectrum from a commercial detector at each image pixel. This is very slow, taking many hours of scanning to acquire even small images. Another common technique is to set spectral energy windows (regions of interest, ROI) and make continuous raster scans. This is fast, but prone to misidentification of elements due to emission line overlaps. Maia provides the best of both methods, i.e. full spectra and fast acquisition, by acquiring data photon-by-photon in streaming mode, and developing elemental maps in real time using a novel analysis method, Dynamic Analysis [1, 2], and using on-the-fly scanning. This requires custom detectors and a novel readout / processing scheme.

2. Architecture

Maia has five major components:

1. A sensor array of 384 diodes, each 1 mm x 1 mm, built on high-resistivity silicon, 400μm thick, arranged symmetrically around a central 2.4 mm diameter hole.
2. An array of low-noise charge amplifiers and pulse shapers, one for each sensor diode [3].
3. A peak detector and de-randomizer circuit which multiplexes the amplifier outputs to a fast Analog-Digital Converter (ADC) for digitization [4].
4. A real-time processor, Hymod[1], which contains a large Field Programmable Gate Array (FPGA), together with a powerful microprocessor system and its associated memories and data interfaces [5].
5. A comprehensive control and analysis software package which implements Dynamic Analysis either off-line in a computer or in real-time in the Hymod parallel processor.

Figure 1 illustrates the general architecture of the system.

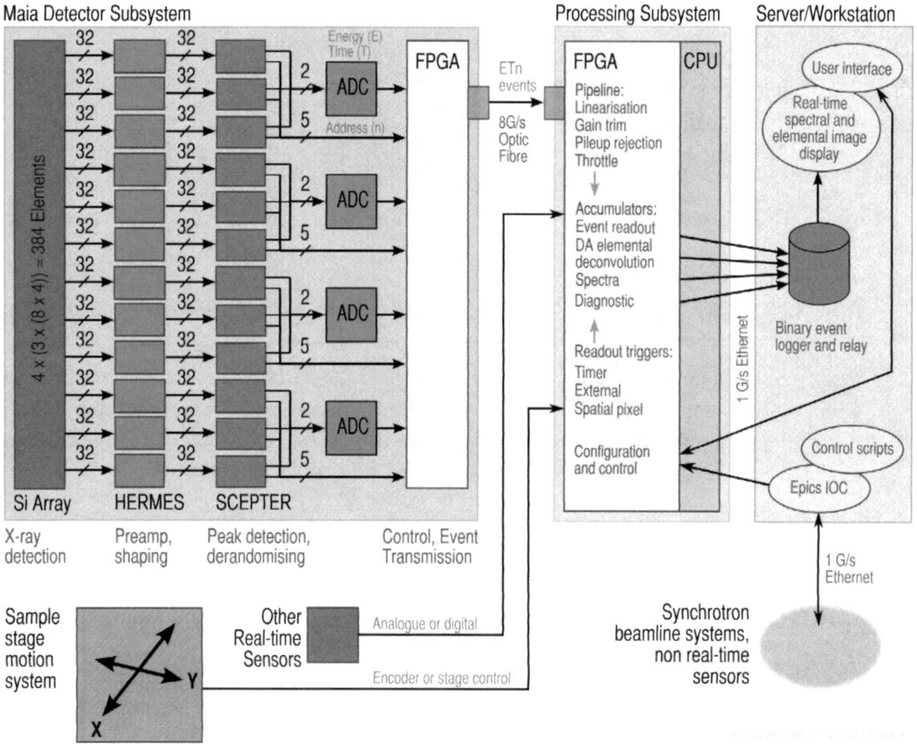

Figure 1. Functional block diagram of the Maia system.

2.1 Diode array

The diode array is fabricated in-house at BNL, using the facilities of the Instrumentation Division Semiconductor Laboratory. It consists of a 20 x 20 array of diodes, each 1 mm x 1 mm, constructed

[1]Hymod: "Hybrid Modular" FPGA-centric processor module.

Journal of Physics: Conference Series **499** (2014) 012001 doi:10.1088/1742-6596/499/1/012001

on a 0.4 mm thick high-resistivity (4-6 kΩ-cm) n-type silicon wafer. The entrance surface was uniformly implanted with phosphorous, and the pixellated side was patterned with boron implants. The implant depth of the phosphorous is roughly 1 μm, and the boron is slightly shallower. Contacts on both sides were formed by depositing aluminum containing 4% silicon to a thickness of 300 nm. The silicon content prevents problems with reactions at the interface leading to destruction of the junction. A cross-sectional schematic of this structure is illustrated in figure 2(a). The central 4 x 4 group of pixels is missing in order to make space for a 2.4 mm diameter hole (and associated guard-ring structures) through which the focused x-ray beam passes. Thus, the total number of detectors is 384. This "backscattering" geometry allows very close approach of the sample to the detector for optimal solid angle, while not restricting the size of the object to be scanned.

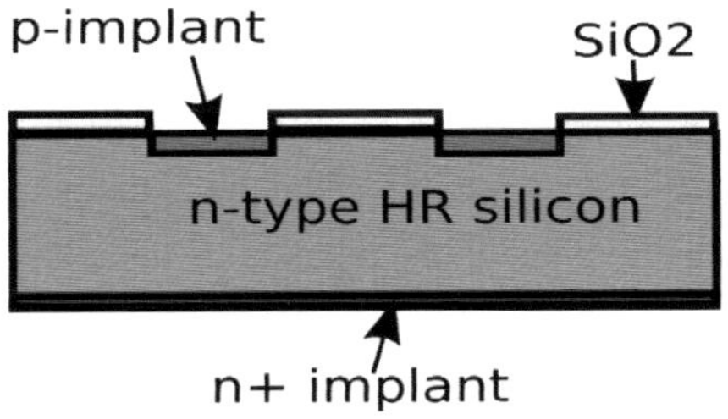

Figure 2(a). Cross-section of high-resistivity silicon photodiode structure (not to scale).

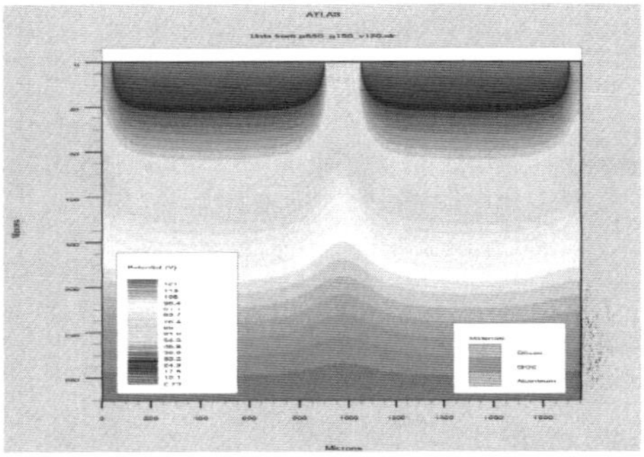

Figure 2(b). TCAD simulation of the potentials around the boundary between two diodes in a monolithic structure. Red= +120 V, violet= 0 V

Assuming that the implanted regions of the diodes are low-field regions and hence constitute dead layers, together with the aluminum contact layer, we would expect good performance of these devices starting at around 1 keV. There are local dips in efficiency due to the absorption edges of silicon and aluminum at 1.84 and 1.56 keV respectively, where it can fall to 30% or so. The vacuum enclosure has a 50 μm thick beryllium window which reaches 50% transmission around 2 keV. At the other end of the energy scale it is useful up to an energy at which the silicon bulk becomes transparent and so inefficient as a detector. Depending on one's criteria, this might be around 16 keV (50% absorption), although the detector is still functional up to 60 keV, at which point the amplifier saturates.

One difficulty with such monolithic devices is the fact that photo-generated charge produced in the bulk, near the boundary between adjacent diodes (up to four), can be split among them, producing what appear to be photons of lower energy simultaneously in each of the pixels. This is referred to as charge sharing, and needs to be considered if the best performance of the detectors is to be obtained. Figure 2(b) shows the potentials inside the bulk under full depletion conditions. It can be seen that the potentials produced provide a barrier for positive charge motion between pixels, but if the charge is produced close to the boundary, lateral thermal diffusion will cause some charge to be directed to a neighbouring diode. This charge division process can occur in any ratio, depending on the details of the sensor structure and the position of the absorption event. For a uniform flood illumination this results in an approximately uniform detected pulse height distribution between the photopeak and zero in each diode. The ratio between the photopeak intensity and the lowest point in this plateau is called the peak-valley ratio. The strength of this low-energy plateau, and hence the peak-valley ratio, depends on the area fraction of the detector which is affected by charge sharing. Thus, larger diodes will have a smaller shared fraction and hence a better peak-valley ratio in general. Our diodes are quite small, and so to maintain a good peak-valley ratio we need to take some additional action.

To solve this problem we chose to add an absorber in front of the entrance surface of the sensor shaped to place all such regions in shadow so that only fully-collected charges will be analyzed. Since we deliberately place the detector very close to the point source, i.e. the focused x-ray spot producing fluorescence from the sample, the shape of this shield is not trivial. It must take account of parallax as all the detectors are off-axis. The solution adopted was to make a layered shield of molybdenum, with three layers, each 0.1 mm thick, having systematically different shapes to shadow the boundary regions even at the edges of the sensor array[6]. Figure 3(a) shows this idea, and figure 3(b) is a photograph of the sensor array fitted with such a shield.

The four silicon tabs seen on the corners of the array serve to provide a good thermal contact with the cold plate on which the sensor is mounted. The cold plate is an aluminum nitride block which is cooled by Peltier devices attached to all four sides. The block is machined to clear the active region of the detector and to provide attachment points for the sensor tabs on each corner.

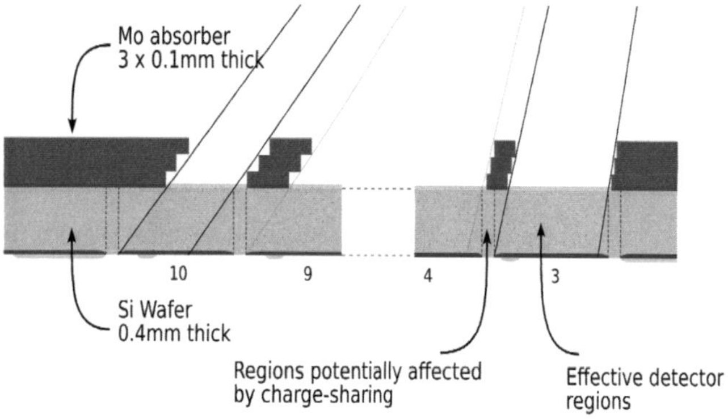

Figure 3(a). The geometry of the layered absorber designed to prevent photon absorption in the regions of the sensor affected by charge-sharing. Reference [6] contains a full description of the design.

Figure 3(b). A photograph of the sensor with its absorber attached. The silicon extensions at each corner serve to provide thermal contact with the Peltier coolers.

2.2 Low-noise amplifiers

Amplification of the small signals produced in the sensor is performed by an Application-Specific Integrated Circuit (ASIC). The chip, HERMES[2], provides 32 channels of very low-noise charge amplification and filtering to produce Gaussian-like pulses. It provides two gain ranges, 0.75 and 1.5 V/fC, and four shaping times in binary steps from 0.5 μs to 4 μs. Channels can be individually disabled in the event of a bad sensor diode, and an external test pulse can be connected to the input of each channel, to allow testing in the absence of a sensor. Since the operation of the input charge amplifier relies on the sensor leakage current to provide its input bias current, the circuit provides for the injection of a 1 pA current on a per-channel basis. This is most often used for system testing

[2]High Energy Resolution Multi-Element Spectrometer

before sensors are bonded, when no leakage current is present. The ASIC also includes pulse-height discriminators and counters for each channel [3], but these are not used in the Maia application.

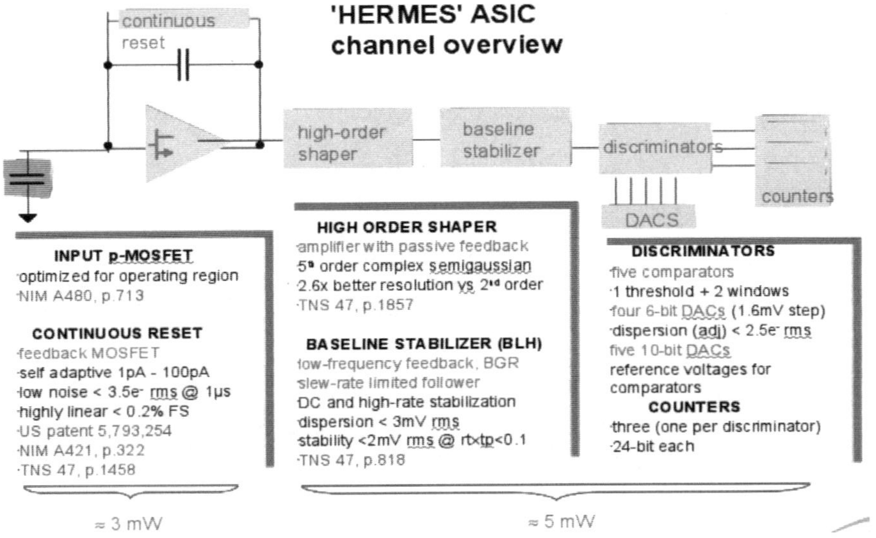

Figure 4. Block diagram of the HERMES ASIC. It contains 32 copies of the circuits shown in the figure [3].

2.3 Peak detector / derandomizer

The Gaussian pulses, which arrive randomly in time, are passed to another ASIC, SCEPTER[3] [4], whose function is to capture the peak height of each pulse and pass it to a fast Analog-Digital converter (ADC). SCEPTER also provides a de-randomization function, allowing efficient use of high-speed ADCs which can only perform conversions synchronously. The chip (see figure 5) contains eight peak-detector modules which are multiplexed among the 32 input channels, providing a regular stream of analog values to the ADCs. The module also provides an analog measure of the time the pulse spends above a defined threshold value, and this is also digitized by a separate ADC and combined with the peak height data for further processing. The time-over-threshold value can be used together with the pulse height information to detect pulse pile-up and reject those events. Reference [4] describes the operation of this ASIC in detail.

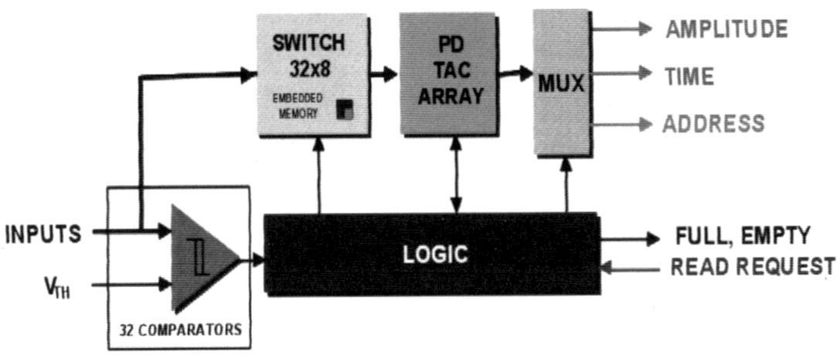

Figure 5. Block diagram of the SCEPTER ASIC. 32 inputs are multiplexed among 8 peak detectors. Routing of inputs to one of the peak detectors is automatic, and simultaneous inputs are correctly handled [4].

[3]Simultaneous Capture of Events with Programmable Timing and Energy Readout

Twelve of each of these ASICs are required to instrument the 384 detectors in the sensor array. The ASICs and sensor are interconnected by 25 μm diameter aluminum wire bonds, as illustrated in figure 6. As can be seen, some of these bonds are extremely long, far outside of the length considered normal for such bonds. It takes some care and skill to form them. Nevertheless, we considered various alternative configurations involving some kind of interposer and we have found that this is the technique which provides the lowest capacitance interconnection, resulting in the lowest noise.

Figure 6. The wire bonds connecting the ASICs to the sensor and to each other. This is the lowest capacitance solution to this interconnection problem, but requires significant skill on the part of the bonder.

The printed circuit board which supports the ASICs is a complex multilayer board with roughly 400 components. A picture of the entire board is shown in figure 7. Some of the components dissipate significant power and, since the board is mounted in an evacuated chamber, there is no convection cooling to remove it. As a result it is necessary to bond the PCB to a water-cooled copper plate. The Peltier elements which cool the sensor to -20 °C also dissipate several Watts of power. The hot side of these elements is also thermally coupled to this plate.

Figure 7. The circuit board mounted on its water-cooled copper heat-sink. This is necessary to maintain a low temperature in a vacuum environment, since there is significant power dissipation from the electronics and also the Peltier coolers. The copper plate is 111 mm x 130 mm, and the active area of the sensor is 20 mm x 20 mm, with the central 4 mm x 4 mm left inactive to accommodate the central hole.

Journal of Physics: Conference Series **499** (2014) 012001 doi:10.1088/1742-6596/499/1/012001

A mechanical model of the detector construction is shown in figure 8(b) [5]. It shows the internal construction. A significant effort was put into minimizing the assembly thickness in order to accommodate the short focal lengths of zone-plate microscopes. As seen in the figure, the beam entrance window is recessed into the vacuum chamber and in operation the zone plate and order-sorting aperture are inserted into this recess to allow its focus to exit onto the sample. The working distance of the sample from the front of the vacuum chamber is around 2 mm. Since the sample translation is parallel to this front face this requirement is not a problem and extended samples can be imaged.

Figure 8(a). A photograph of the complete detector head. The lower module contains the detector module shown in figure 7, while the upper module contains the digital interface to the Hymod. As a scale reference, the holes in the optical table are 25 mm apart.

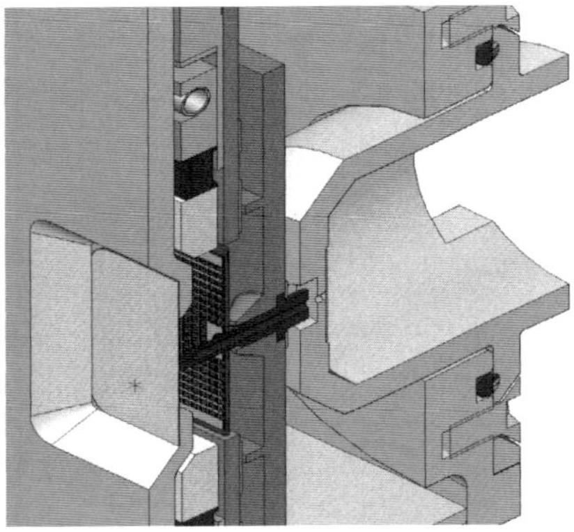

Figure 8(b). Computer model cross-section of the detector mechanics. It shows the re-entrant entrance cavity to take a zone plate assembly, and the molybdenum tube which shields the sensor from radiation scattered by either entrance or exit beryllium windows [5].

2.4 Real-time processor

The digitized values of peak-height and time-over-threshold are combined with several other values and passed to the Hymod module. This device consists of a large Field-Programmable Gate Array (FPGA) which performs high-speed computation on the data, closely coupled to a fast microprocessor and its associated memories and interfaces. Since each detector and its analog processing chain is slightly different, the system must keep track of which photon event came from which detector so that appropriate energy calibration can be performed. It must also associate each photon with a particular sample position in order to reconstruct the scanned image. Thus, each photon generates a "photon event", which contains the following information:

Peak height
Time-over-threshold
Detector number

This information is encoded into a 32-bit packet and sent for processing. Other events are generated by the system and merged into the data stream. A change in the sample stage coordinates, for example, is such an additional event type. All packets are 32 bits, and contain a tag identifying what type of event it is. Processing can consist of simply logging the data to high-speed disks, or immediate processing in the FPGA, or both. The FPGA processing generates elemental maps photon-by-photon, in real time, such that elemental maps are produced on-screen as the scan progresses. This is a major advance, since it allows the user to see very quickly if she is getting the data she wants, and if not, to stop the scan and make adjustments. The system does much more than this, and the reader is referred to reference [5] for further details. A flow-chart of the processing is shown in figure 9.

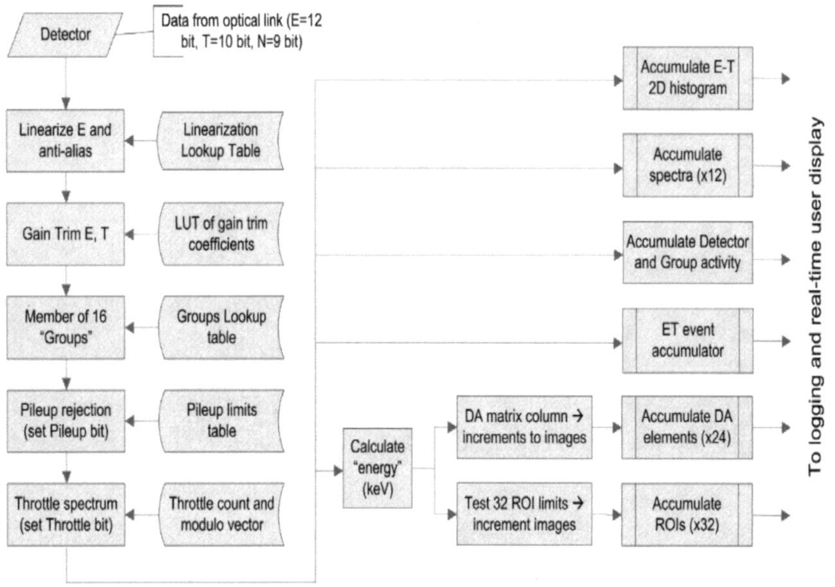

Figure 9. A flow-chart of the processing pipeline performed within Hymod. In the diagram, E refers to photon energy, T refers to the time-over-threshold value (used to detect pile-up) and N refers to the index of the detector which collected the photon. Hymod is capable of handling up to 50 million events per second [8].

A comprehensive software package has been written to provide the User Interface for the detector.

3. Future directions

Our primary development goals aim to improve both energy resolution and throughput. Both will require significant effort, both in ASIC design and novel sensor structures. A new ASIC is currently under test, and promises to markedly reduce the contribution of electronic noise to the overall system noise level. It does this by improvements to the preamplifier design, removing all of the counting logic, which is not used by Maia, and by making all digital signals differential so that noise from fast digital signal edges is greatly reduced. Many of the circuits used are based on those used in earlier ASICs developed at BNL [7]. We are also studying alternative sensor structures with a view to reducing their capacitance, which is a significant component of the noise budget for the current Maia design. In general, signals from lower-capacitance sensors can be processed faster, leading to higher throughput. Silicon Drift Detectors (SDDs) are an obvious, though complex choice. However, other, simpler solutions may exist and we are investigating them. In order to accommodate any eventuality, the new ASIC can accept either electron or hole signals. In the more distant future, we are looking at moving from the wire-bonding interconnect scheme to a solder bump bonding process. This would serve two functions; it would reduce the risk of failure of the bonding process and it would reduce the footprint of the sensor-ASIC block. Against those advantages, there is the disadvantage that the power dissipated in the ASICs is in close thermal contact with the sensor, and so more cooling power would be needed.

4. Conclusions

We have developed a detector system specifically designed for x-ray fluorescence microprobes. We took the non-traditional step of working in a backscattering geometry, rather than at 90 degrees to the incident beam. This makes possible a reduction in the sample-to-detector distance to a few millimeters without restricting the scan range of the sample. Samples of order 1 m x 1 m have been scanned. Contrary to popular belief, Compton scattering is not overwhelming. It can in fact provide a useful estimate of sample density and can be used for imaging, for example to provide registration between metal images and surrounding organic material such as is found in a biological organism. The data acquisition system is also non-traditional, borrowing as it does from high-energy physics practice. It streams data continuously, event-by-event, rather than as complete spectra. Ryan's Dynamic Analysis method [2] takes advantage of this architecture to allow real-time analysis of elemental contributions as data collection proceeds.

The Maia system has been used for a wide range of applications. The papers by Ryan et al [8] in these proceedings and by Lombi et al [9, 10, 11] describe some of them in detail. It has enabled a new mindset for fluorescence elemental mapping, where large images with high resolution (high-definition imaging) are acquired, rather than making a low-resolution image first in order to decide which regions of the sample merit finer scans. This earlier method can easily result in missed features due to its inevitable under-sampling. If the coarse scans are made using a larger probe beam to avoid under-sampling, small particles are effectively diluted and may also be missed. The speed of acquisition also enables other modalities involving scans in additional dimensions, e.g. fluorescence tomography (e.g. [12]) or XANES imaging (e.g. [13]) .

5. Acknowledgments

This work was supported at BNL by the U.S. Department of Energy, Office of Science, Office of Basic Energy Sciences, under Contract No. DE-AC02-98CH10886. The CSIRO work took place within the CSIRO Sensors and Sensor Networks Transformational Capability Platform. NJIT authors were supported by the National Science Foundation, Major Research Instrumentation Grant DMR 0722730. SLAC is supported by the U.S. Department of Energy under contract number DE-AC02-76SF00515.

References

[1] Ryan C G, Siddons D P, Kirkham R, Dunn P A, Kuczewski A, Moorhead G, De Geronimo G, Paterson D J, de Jonge M D, Hough R M, Lintern M J, Howard D L, Kappen P and Cleverley J The New Maia Detector System: Methods for High Definition Trace Element Imaging of Natural Material 2010 X-Ray Optics and Microanalysis, AIP Conference Proc. **1221** 9-17

[2] Ryan C G 2000 International Journal of Imaging Systems and Technology (special issue on “Advances in Quantitative Image Analysis”) **11** 219-230

[3] De Geronimo G, O'Connor P, Beuttenmuller R H, Li Z, Kuczewski A J and Siddons D P 2003 IEEE Transactions on Nuclear Science **50** 885-891

[4] Dragone A, De Geronimo G, Fried J, Kandasamy A, O'Connor P and Vernon E 2005 IEEE Nuclear Science Symposium Conference Record **2,** 914-918, 2013

[5] Kirkham R, Dunn P A, Kucziewski A, Siddons D P, Dodanwela R, Moorhead G, Ryan C G, De Geronimo G, Beuttenmuller R, Pinelli D *et al.* 2010 *AIP Conf. Proc.* **1234** 240-243

[6] Ryan C G, Kirkham R, Hough R M, Moorhead G, Siddons D P, de Jonge M D, Paterson D J, De Geronimo G, Howard D L and Cleverley J S 2010 *Nucl. Instr. Meth A* **619** 37-43
[7] G De Geronimo, P Rehak, K Ackley, G Carini, W Chen, J Fried, J Keister, S Li, Z Li, D A Pinelli, D P Siddons, E Vernon, J A Gaskin, B D Ramsey, and T A Tyson. IEEE Transactions on Nuclear Science, vol. **57**, no..3, June 2010, 1653-1664.
[8] C G Ryan, D P Siddons, R Kirkham et al., these proceedings.
[9] Lombi E, de Jonge M D, Donner E, Ryan C G, Paterson D 2011 Anal. Bioanal. Chem. 400 1637-1644 .
[10] Lombi E, de Jonge M D, Donner E, Kopittke P M, Howard D L, Kirkham R, Ryan C G and Paterson D 2011 PLoS One (Public Library of Science) 6(6) e20626.
[11] Lombi E, Smith E, Hansen T H, Paterson D, de Jonge M D, Howard D L, Persson D P, Husted S, Ryan C and Schjoerring J K 2011 J Experimental Botany 62 273–282
[12] Carey A-M, Lombi E. Donner E. de Jonge M D, Punshon T, Jackson B P, Guerinot M L, Price A H, Meharg A A 2012 Analytical and Bioanalytical Chemistry 402 3275-3286
[13] Lintern M, Hough R and Ryan C 2012 J. Geochem. Explor. 112 189-205

Maia X-ray fluorescence imaging: Capturing detail in complex natural samples

C G Ryan[1], D P Siddons[2], R Kirkham[3], Z Y Li[2], M D de Jonge[4], D J Paterson[4], A Kuczewski[2], D L Howard[4], P A Dunn[3], G Falkenberg[5], U Boesenberg[5], G De Geronimo[2], L A Fisher[1], A Halfpenny[1], M J Lintern[1], E Lombi[6], K A Dyl[7], M Jensen[3], G F Moorhead[3], J S Cleverley[1], R M Hough[1], B Godel[1], S J Barnes[1], S A James[4], K M Spiers[4], M Alfeld[5], G Wellenreuther[8], Z Vukmanovic[7,1] and S Borg[1]

[1] CSIRO Earth Sciences and Resource Engineering, 26 Dick Perry Avenue, Kensington, WA 6151, Australia
[2] Brookhaven National Laboratory, Brookhaven, Upton NY, USA
[3] CSIRO Materials Science and Engineering, Normanby Road, Clayton VIC 3168, Australia
[4] Australian Synchrotron, Blackburn Road, Clayton VIC 3168, Australia
[5] Photon Science, DESY, Notkestr. 85, D-22603 Hamburg, Germany
[6] Centre for Environmental Risk Assessment and Remediation, University of South Australia, Building X, Mawson Lakes Campus, Mawson Lakes SA 5095, Australia
[7] Department of Applied Geology, Curtin University, GPO Box U1987, Perth, WA 6845, Australia
[8] European XFEL, Albert-Einstein-Ring 19, 22761 Hamburg, Germany.

E-mail: chris.ryan@csiro.au

Abstract. Motivated by the challenge of capturing complex hierarchical chemical detail in natural material from a wide range of applications, the Maia detector array and integrated real-time processor have been developed to acquire X-ray fluorescence images using X-ray Fluorescence Microscopy (XFM). Maia has been deployed initially at the XFM beamline at the Australian Synchrotron and more recently, demonstrating improvements in energy resolution, at the P06 beamline at Petra III in Germany. Maia captures fine detail in element images beyond 100 M pixels. It combines a large solid-angle annular energy-dispersive 384 detector array, stage encoder and flux counter inputs and dedicated FPGA-based real-time event processor with embedded spectral deconvolution. This enables high definition imaging and enhanced trace element sensitivity to capture complex trace element textures and place them in a detailed spatial context. Maia hardware and software methods provide per pixel correction for dwell, beam flux variation, dead-time and pileup, as well as off-line parallel processing for enhanced throughput. Methods have been developed for real-time display of deconvoluted SXRF element images, depth mapping of rare particles and the acquisition of 3D datasets for fluorescence tomography and XANES imaging using a spectral deconvolution method that tracks beam energy variation.

1. Introduction

The design of the Maia detector array and imaging system has been motivated by the need to capture intricate spatial and chemical detail in complex natural samples. While the motivation for ever higher spatial resolution in the development of X-ray microscopy is driven by the need to capture the finest structures, these details need to be interpreted within their larger spatial context. For instance, evidence for diffusion, reaction or crystallization along grain boundaries provides geologists with clues to process and timing. However, these textural and chemical clues must be seen within the context that the hosting structure provides. For example, areas which show evidence for infiltrating fluids and alteration may occur from the scale of inter-grain boundaries up to crustal scales spanning 12 orders of magnitude. If we concentrate on the goal of maximizing information content within the scale of geological samples presented for X-ray fluorescence mapping analysis (e.g. in the form of mineralogical thin-sections) then the textural clues are contained within spatial scales from a few cm down to the focussed X-ray beam size of ~1 μm. This presents a challenge of capturing detail over ~4 orders of spatial scale.

An example of a common fine detail problem can be described as "needles in a haystack". These are very rare particles or mineral phases, which are minute in size and may occur only at a rate of a few per square centimetre and challenge the spatial resolution of the X-ray microprobe. In geology, examples include rare gold particles sought in studies of ore genesis, rare resilient platinum group minerals that may survive in the Earth's mantle for billions of years, elusive gold precipitates in plants growing over ore deposits or rare pre-solar condensates that may provide clues to early solar system formation if they can be identified. However, these objects may occur only at average concentrations of a few ppb, concentrated in a few sub-micron phases, located at unknown positions somewhere within the thin-section. X-ray fluorescence microscopy (XFM) is ideal for this type of analysis as it can detect and image a sub-micron particle at depth within a section.

Capturing detail over 4 orders of magnitude of spatial scale demands the ability to produce images of $\sim 10^8$ pixels in practical time-scales, such as 3–10 hours (e.g. 0.1 to 0.4 ms per pixel). Scanning at this pixel rate demands an approach that eliminates read-out overheads. To obtain adequate counting statistics within each of these pixels at least for major/minor components (e.g. $\sim 10^3$ counts) demands total detector count-rates of ~10 M/s. The approach taken with the Maia detector has been to spread such high count-rates over a large number of detectors in an array, rather than attempt heroic count-rates in a small number of detectors with its inherent risk of serious pileup artefacts and degraded energy resolution. This has only been possible by making use of the large scale analogue circuit integration developments at Brookhaven National Laboratory [1].

A parallel goal of the Maia development has been to achieve high count-rates by maximizing total detector array solid-angle rather than increasing flux. In this way sample damage can be minimized. This has entailed design of Maia coordinated with the design of the XFM end-stations [2] at the Australian Synchrotron (AS). Its KB mirror end-station, with its large working distance of 69 mm provided little challenge. However, the coordinated design of Maia and its Xradia nanofocussing end-station demanded several design iterations to accommodate the 25 mm working distance of its diffractive optics. The Maia detector array uses an annular, "back-scattering" geometry, where the beam passes through the detector located upstream of the sample. The annular configuration enables a large solid-angle without imposing severe restrictions on sample size or the scale of scanning.

The challenge of data acquisition at count-rates up to or exceeding 10 M/s, with pixel times as short as 0.1 ms or less while avoiding readout overheads, was tackled by applying a nuclear physics approach to data acquisition using event-mode data collection with real-time processing. In addition, the real-time processing capability provides a platform in which established techniques for real-time spectral deconvolution can be applied, which paves the way for the display of deconvoluted element images in real-time.

This paper outlines the design and performance of the Maia detector array and imaging system as implemented for the XFM beamline at the AS in a coordinated endstation-detector design exercise. This paper will also present new improvements in Maia technology that are benefitting XFM analysis at the P06 beamline [3] at Petra III in Germany, which uses a new version of Maia with improved energy resolution. It also provides a discussion of the advantages and drawbacks of the annular geometry.

2. The Maia detector array and imaging system

The detector array comprises a 20 × 20 array of 1 mm^2 individual detectors with a central group (4 x 4) removed (figure 1) to accommodate the central Mo collimating tube for the passage of the X-ray beam through the detector [1,4]. A Mo mask grid over the array absorbs X-rays that may otherwise result in charge sharing between detectors [5]. A target-detector distance of 10 mm provides a total solid angle of 1.3 sr. Each detector is directly wire-bonded onto the inputs of 12 HERMES 32-channel custom applications specific integrated circuits (ASICs) providing pre-amplification, baseline restoration and 5-pole quasi-Gaussian pulse shaping to each detector channel [1,6]. Each HERMES channel drives an input of 12 SCEPTER 32-channel de-randomizer ASICs, which sample pulse-height (E) and time over threshold (T) for each detector pulse and provide a sample-and-hold function for digitization [7], multiplexed into 4 banks of high-speed 14-bit synchronous ADC pairs. A field-programmable gate-array (FPGA) in the detector head packages detected events into data representing events {n, E, T}, for detector channels 'n', and transmits them over a dedicated 8 Gb/s quad-channel fibre optic link to the real-time processing sub-system located outside the beamline enclosure [4].

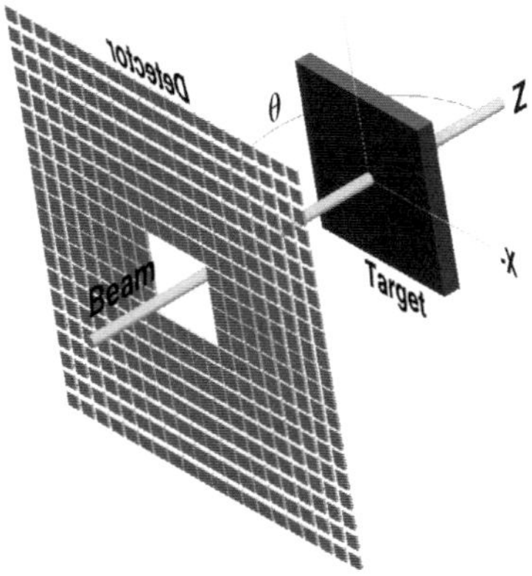

Figure 1 The Maia detector annular array comprises a 20 x 20 array of 1 mm^2 silicon detector elements, with a central group (4 x 4) removed to accommodate the passage of the X-ray beam (see detector fabrication details in ref: [1]). The plane of the array is 10 mm from the focussed spot on target. The size of the detectors depicted here in perspective, which diminish towards the outer perimeter, reflects the effective size after masking to avoid charge-sharing between detectors (the mask is detailed in ref: [5]).

The real-time processing sub-system uses a CSIRO HYMOD processor. It comprises a FPGA, which provides pipelined parallel processing of the event stream, and a co-processor to handle control and display client requests and event transmission over Ethernet [4]. A new event can be clocked through the FPGA each clock cycle at up to 50 M/s. In parallel, the HYMOD system samples pulses from the upstream and downstream ion-chamber V-to-F units to count beam flux on a per pixel basis and tracks stage encoder position to enable each output event record to be tagged by stage position. The event processing pipeline comprises: (i) ±0.5 dither applied to E to reduce subsequent aliasing effects, (ii) linearization correction, (iii) linear gain trimming to match E and T across detector channels, (iv) energy calibration mapping, (v) pile-up rejection, (vi) individual detector count-rate monitoring, (vii) dead-time and pileup monitoring across the array, and (viii) real-time element image accumulation triggered directly by sample stage motion. Processed events, optionally sub-sampled ('throttled') on the basis of energy, are also logged to disk, tagged by sample stage pixel coordinates with incident and transmitted flux counts accumulated in that pixel. Throttling reduces the size of raw data files through suppression of strong element lines or scatter peaks.

3. Treating spectral complexity

Many applications, especially those in the environmental, planetary and geological sciences, are characterized by complex synchrotron XRF spectral content with numerous overlaps between elements. The analysis of complex SXRF spectra is well developed, typically using a least-squares fitting approach with X-ray line yields based on a detailed model of the sample (e.g. [8,9]). The Dynamic Analysis (DA) method is based on an extension of this approach to imaging. In the limit of a linear least-squares fit, where non-linear terms for peak width, tailing and energy calibration adjustment have been constrained by prior measurement (or prior non-linear fit iterations), the solution

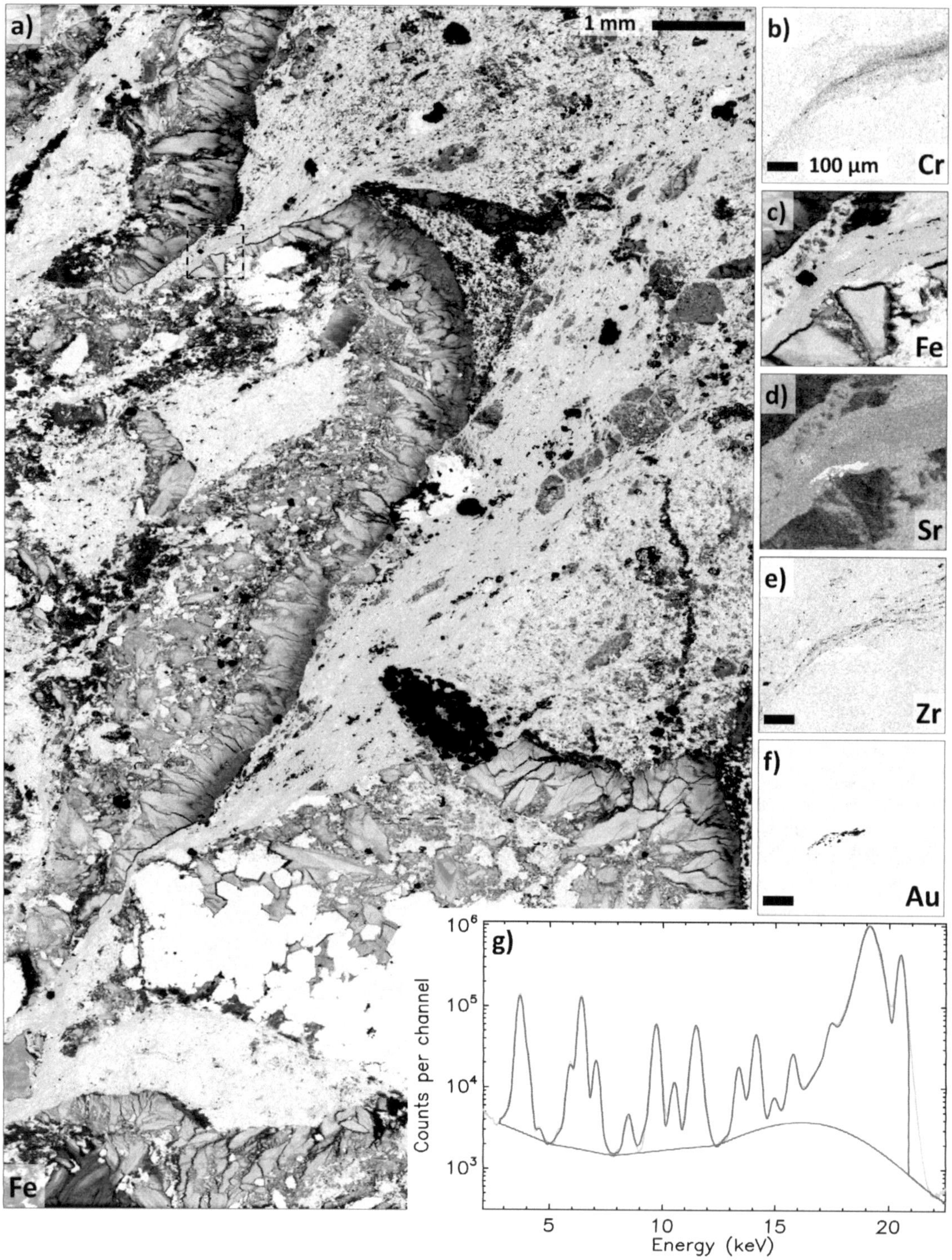

Figure 2 Maia images of a thin-section from the western shear zone of the Sunrise Dam deposit, Western Australia acquired on the P06 beamline; a) large area Fe image (4050 x 8000 pixels, 8 x 16 mm^2) showing a zoom in area (box top left); images (0.61 x 0.70 mm^2) from area in (a) for b) Cr, c) Fe, d) Sr, e) Zr and f) Au; and g) total spectrum for detector #124 for the central portion of the scan (overlaid by GeoPIXE fit and background), for total detector count-rate of 4 M/s (τ = 1 μs).

of the fitting problem using the function linearization method [10] reduces to the solution of a set of simultaneous equations. In matrix form, these equations can be rearranged into the form of element intensities given by a matrix transformation of the spectrum of interest [11].

Using a fundamental parameters approach to SXRF analysis, the element intensities can be related to concentration, which results in a matrix transformation directly from input spectrum to element concentration vector [12]. In the Maia case, the model includes the effects of the Mo mask on individual detector solid-angles and the changing spectral relative intensities across the array, as an extension of the layered sample model in the GeoPIXE software [5,13,14]. The model produces a set of relative sensitivity factors (RSFs) for the individual detectors across the array reflecting changing distance, angle, absorption paths and intrinsic efficiency. Application of the RSF approach enables any sub-set of detectors, or individual detectors, to be used in a particular analysis. An example is the mapping of particle depth, which makes use of contrasting self absorption in the sample experienced by radial groupings of inner and outer detector elements [15].

The DA method can be used for rapid spectrum processing [11,15,16]. However, this approach also can be used to map each event in turn onto its element concentration contributions [17]. Some of these contributions are negative, which is how overlap and background subtraction are accomplished [12]. In the Maia detector processor, these contributions are accumulated for all events in the current image pixel in what reduces to a simple lookup-accumulate operation implemented in the HYMOD FPGA [14]. When the stage moves out of the pixel, the accumulator is read-out as a concentration record, which is displayed immediately by the real-time display client.

4. Application methods using Maia

The key capabilities of the Maia detector system that enable the high throughput of the XFM beamline are (i) transit times per pixel (referred to below as "dwell time") down to 50 μs [14], (ii) close detection geometry providing a solid angle of 1.3 sr [5], (iii) high sustained count rates, normally up to 12 M/s, (iv) 100×100 mm^2 scan range and image area (600×300 mm^2 on the XFM large format stage), (v) image pixel count beyond 10^8 pixels, (vi) real-time spectral deconvolution and image display, and (vii) moderate inelastic scattering signal that provides good collective images of light elements [18], especially useful for biological samples [19-22]. Typically, pixel times and image areas are not generally a limitation, and these are chosen to suit a particular application. This has enabled a range of application methods at the XFM beamline and most are finding application at the P06 beamline at Petra III.

4.1. Large area mapping

Scans of complex natural samples can be performed over 100–200 mm^2 at full spatial resolution of 2 μm on the XFM KB endstation. Larger areas to 10 cm^2 are typically done at 4–6 μm resolution. Typical image size is 10–100 M pixels acquired over 2–10 hours [14]; the largest image to date is 620 M pixels. Using a collimated beam, larger objects, such as works of art, have been imaged at 50–100 μm resolution [18] over areas up to 600×300 mm^2. This method has been applied to studies in the earth [23-30], planetary [31], environmental [32,33], medical [19,34], biological [20-22,35-41] and material [42] sciences as well as cultural heritage [18,43]. A similar capability has been established on the P06 beamline (e.g. figure 2).

4.2. Survey and detail mapping

Large area mapping can be performed at small dwell time per pixel to provide overview survey images (e.g. 6 M pixels in 1 hour) in major and minor elements from which smaller detail areas are selected for scanning at increased dwell time and/or spatial resolution [24,33,15]. Real-time imaging enables these selections to be made during survey image acquisition.

4.3. Depth mapping

The Maia detector array subtends a large range of takeoff angles for X-rays exiting the sample surface, with the angle between the sample normal and ray to the centre of individual detectors varying between 13.9° and 52.6°. This makes the outer detectors much more sensitive to self-absorption for X-rays from a particle at depth in a sample than the inner detectors. Such a difference can be harnessed to provide an imaging contrast based on particle depth and also a quantitative measure of depth for discrete identified particles [15]. This approach has been used to locate rare gold precipitates in eucalyptus leaves [29] (e.g. figure 3) and rare phases including platinum group minerals in mantle derived rocks and meteorites [15].

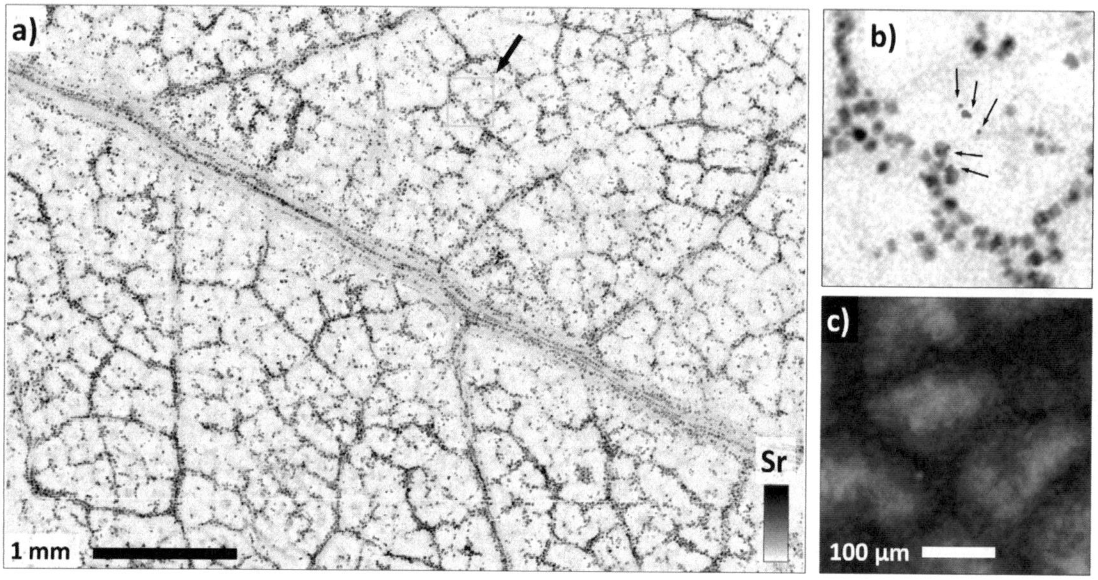

Figure 3 Maia XFM images of a portion of Eucalyptus leaf showing a) Sr distribution (5.6 x 3.8 mm^2, 2814 x 1894 pixels) showing oxalate crystals (Sr is a proxy for Ca), b) zoom in on Sr map (0.36 x 0.36 mm^2, from box in (a)) showing Au grains (red) indicated by arrows (two attached to oxalate crystals), and c) Compton scattering map (see ref: [29]).

4.4. XANES image stacks and XANES tomography

The capability of the Maia detector for imaging to around 10^8 pixels paves the way for 3D imaging modalities, such as XANES imaging, where the production of images up to $\sim 10^6$ pixels can be repeated $\sim 10^2$ times with changing beam energy through an absorption edge from which chemical state images can be projected [25,43,45]. Repeated single slice tomography at energies across an absorption edge also can be used to provide cross-sectional chemical state images [48].

4.4.1. *DA method for XANES imaging.* The DA method provides a linear least squares fit result at real-time processing speed. The model includes all excited X-ray lines, a background term and elastic and inelastic scattering peaks, where inelastic scattering is treated using an empirical function [44]. The method has been extended to XANES imaging by shifting the centroids of the elastic and inelastic peaks to track the changing beam energy. The full non-linear fit is performed on the SXRF spectrum at the top beam energy for the XANES scan, in which the shape of the background and the inelastic scattering distributions are determined. Then a series of DA matrices are generated as the beam energy is reduced according to the method of Ryan *et al.* [12,14] with the elastic and inelastic peaks shifted to track the change in beam energy. During processing of the event stream, the lookup table operation first selects the correct DA matrix for the beam energy and then selects the matrix column based on event X-ray energy to increment a 2D result array indexed by XY position repeated over beam energy.

The benefit of this approach are XANES spectra for each XY with subtraction of contributions due to scattering, background and other elements, including those experiencing their own edge structure within the XANES beam energy range [45]. This method is part of the GeoPIXE software suite [46].

4.5. 3D fluorescence tomography

Large pixel count can be recast as a large voxel count for 3D fluorescence tomography, particularly where matrix absorption is low. 5 μm tomographic data have been acquired using a KB mirror pair with ~2 μm resolution over ~1 mm specimens [37]. Single-slice data can be obtained from highly radiation-sensitive specimens, such as fresh and fully hydrated plant roots, in a matter of minutes [20,21,38,39]. Measurement of the Ge distribution in an intact rice grain with 2 μm voxels over 2–3 mm dimensions utilized high-definition 2D fluorescence tomography, with 2001 projections and a measurement time of the order of 3 hours [22] (figure 4a). Maia has enabled fluorescence tomography to become routine; however the ability to measure ever larger specimens now demands a general solution to the self absorption problem (see ref: [47] and references therein).

5. Maia detector performance

Recent tests of the Maia system showed that the measurements of the RSFs across the array for thin standard metal foils [49] show agreement with the model values at better than 3% relative standard deviation [50]. Further tests, aimed at demonstrating quantitative element imaging over a wide count-rate range, used these metal foils and a wide range of beam flux to produce total count-rates from 0.1 to 12 M/s, which corresponds to ~0.1 to ~10-40 k/s in individual detectors. Pileup rejection used the E-T behaviour to identify and reject pileup events, as outlined by Dragone *et al.* [7], and each event T provided a measure of dead-time. Accounting for pile-up rejection and dead-time losses enabled quantitative analysis of the metal standards using GeoPIXE consistent to within a relative standard deviation of 0.2% for count-rates from 0.1 to 12 M/s [50]. Interestingly, total dead-time was only 5% at a total count-rate of 12 M/s (1 μs peaking time τ). Maximum count-rate is limited at present by the selected SCEPTER clock to 36 M/s.

Ongoing development of the Maia detector has yielded improvements in the energy resolution and aided the detection of lower energy X-rays. The new model 384B, tested recently at the P06 beamline at Petra III, achieved an energy resolution for Mn Kα with an average FWHM of 240 eV across the array, using a HERMES peaking time of 4 μs, and enabled X-ray detection down to 2.0 keV. Figure 2g shows a typical Maia detector spectrum at a total count-rate of 4 M/s (with overlaid GeoPIXE fit and SNIP background [51]). Normalization to flux counts on a pixel by pixel basis is now routine, which helps to eliminate the effects of pixel dwell fluctuations caused by scanning stage motor speed variation.

6. The annular geometry

The annular configuration enables a large detection solid-angle without imposing restrictions on the size of samples or the scale of scanning. This approach enables scanning of virtually any sample size while maintaining large solid-angle. Sample areas as large as ~0.6 × 0.3 m^2 (e.g. works of art) can be accommodated at the XFM beamline [18]. The cost of this approach is increased scatter compared to a small solid-angle detector located at 90° along the direction of typical horizontal beam polarization. As shown in figure 4c, departure from 90° results in at least a 30-fold increase in background for a small solid-angle silicon drift detector (SDD) [52]. Spectral background at 90° remains limited by the combined effects of multiple scattering in the sample and environment as well as incomplete charge collection background in the detector.

However, pursuing large solid-angle tends to wash out the advantage of the 90° geometry, with only a marginal penalty for choosing the annular geometry (θ=180°); scatter increases by less than a factor of 3 for Maia between 90° and 180° (figure 4b). The great increase in detector solid-angle afforded by the close annular geometry more than compensates for this slight disadvantage allowing a 200-fold increase in counting statistics in this example, which yields improved detection limits and is crucial for obtaining the high level of counting statistics needed for quality imaging. Figure 5 illustrates that large detector solid angle helps avoid inadequate counting statistics, which can degrade effective image spatial resolution well beyond actual beam resolution.

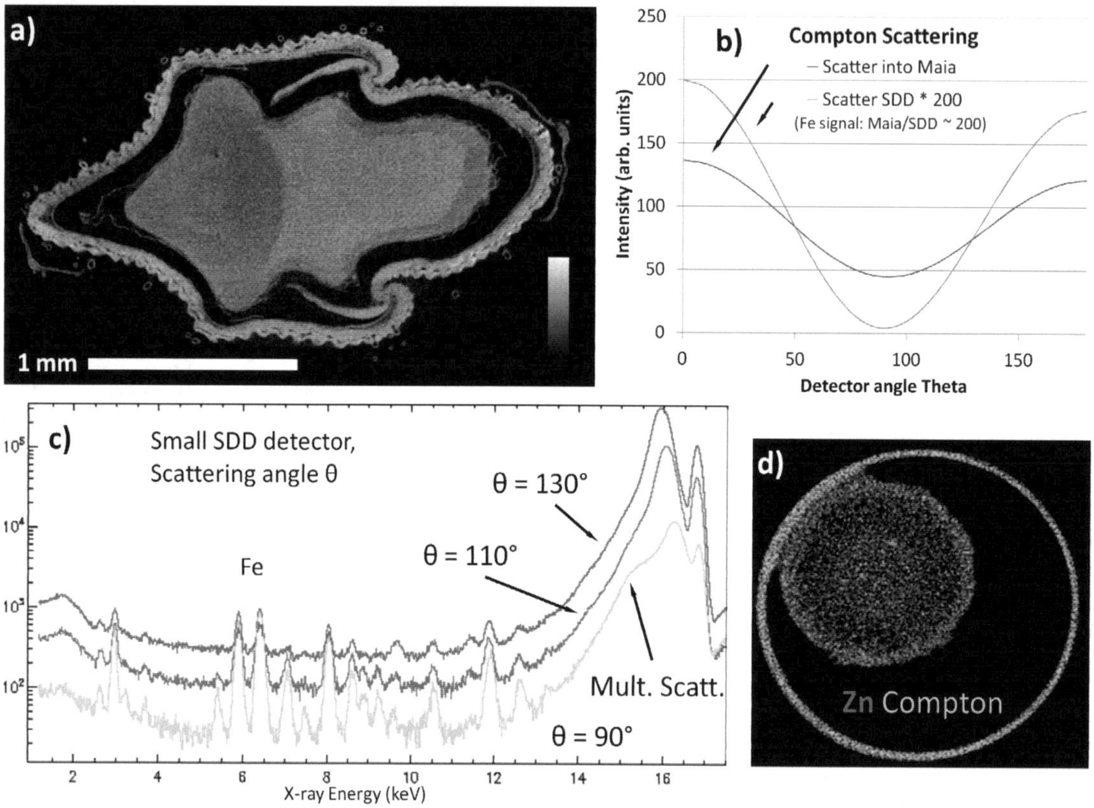

Figure 4 a) Maia Compton scattering tomography data reconstructed to show a section through a rice grain (4.5 mm scan {2291 pixels} at 2001 projections, see ref: [22]); b) angular distribution of Compton scattering for a small SDD detector (Ω=6.2 msr) ignoring multiple scattering (green) and the Compton angular distribution using a Maia detector (Ω=1.3 sr, scaled down x200 in red) as a function of detector angle theta (annular geometry at 180°); c) spectra for a thin NBS 1832 standard normalized to fluence using a small SDD detector as a function of detector angle θ [52], and d) reconstruction of Maia fluorescence tomography data for a section through a capillary tube containing a rootlet (see ref: [21]) showing the tube and root imaged using Compton scattering (green).

A final point in this discussion is the observation that the scatter signal has proven very useful in its own right as a means to image the structure of biological samples as a collective proxy for light elements. Compton scattering from light elements is dominant in biological samples, illustrated in figures 4a,d and 3c. Numerous examples of Maia imaging of light elements using scattering can be found elsewhere (e.g. [19-22,40]).

7. Application examples

The Sunrise Dam Gold Mine contains ~14 M oz gold and is the largest known deposit in the Laverton Tectonic Zone of the Eastern Goldfields province, Western Australia [53]. Gold is hosted in shear-parallel flat veins, steep stockwork veins and hydrothermal breccia bodies. The veins typically have quartz-carbonate mineralogy with sericite-sulphide alteration of the surrounding wall rock. The majority of gold is hosted within the veins with little evidence for gold within the wall rock [53]. Maia

XFM maps of a deformed Au-bearing quartz-carbonate vein from the western shear zone of Sunrise Dam (figure 2) shows evidence for a complex deformation and alteration history associated with multiple Au populations [54]. Chemical zonations within the carbonate minerals indicate multiple phases of vein growth and subsequent deformation and alteration [55]. Coarse Au is hosted within the fractured vein, while a rare, finely textured Au population is hosted in sericite altered shear zones on the boundary between the vein and the wall rock (figure 2f). These sheared, sericite rich zones, are characterised by abundant hydrothermal zircon and an elevated Cr content (figure 2b, indicating presence of fuchsite, a common alteration phase at Sunrise Dam). The contrast in size between the Au populations in the vein and sheared boundary between the vein and wall rock may indicate multiple Au deposition events, or (more likely) the gold grain size is controlled by the rheology of the host phases and space available for growth.

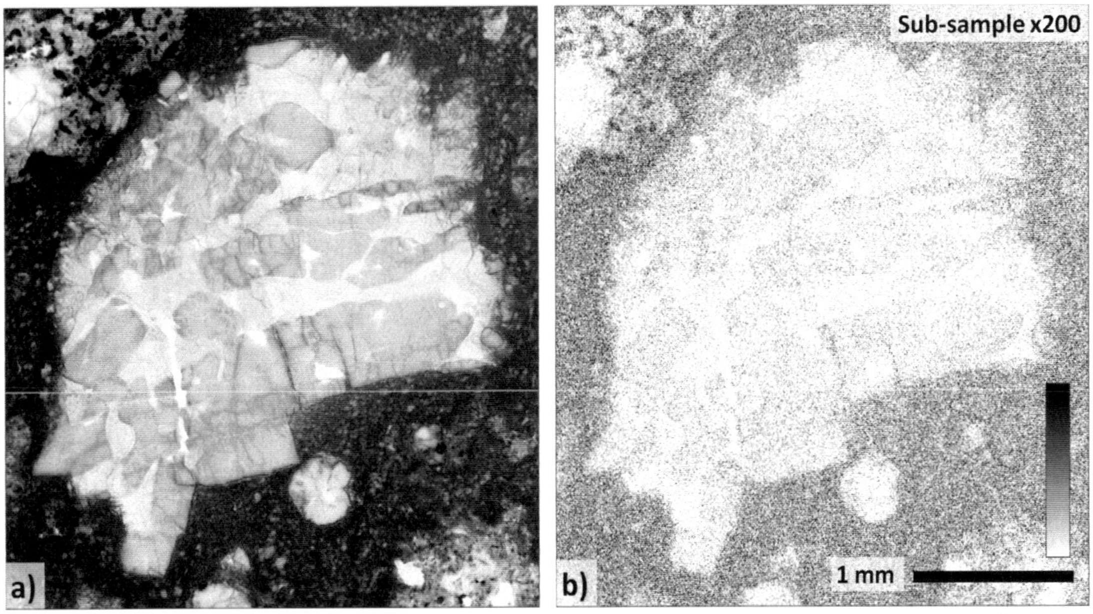

Figure 5 a) Maia Fe image of a single chondrule in an Allende meteorite sample ([31], see description in ref: [15]); and b) the same data-set sub-sampled x200 to illustrate how lower counting statistics degrades image quality, perceived detail and effective spatial resolution.

Geochemical sampling of vegetation above known gold ore bodies has been shown by Lintern [56] to exhibit anomalously high Au concentrations. Hence, the uptake of metals by vegetation has potential as a mineral exploration technique. In arid regions of Australia, species of Eucalyptus and Acacia can develop root systems that tap water and dissolved metals down to depths of ~40 m. However, the mechanism for Au enrichment in vegetation is unknown and it has been difficult to distinguish natural uptake from surface contamination. A recent study by Lintern *et al.* [29] combined field sampling of fresh Eucalyptus leaves and Maia XFM imaging to locate and image rare gold precipitates located within the leaves. The Freddo gold prospect, Western Australia, was selected as a site to investigate the potential mechanisms for the chemical transport of Au and uptake by plants with the results compared with laboratory studies using seedlings of native Eucalyptus and Acacia dosed with Au. The Freddo deposit is undisturbed by mining activity and has no known sources of surface contamination. The sub-economic ore body is concealed beneath ~30 m of barren overburden. Nevertheless, gold precipitates were located using XFM in numerous leaf samples; the speed of Maia imaging enabled 60 leaf discs samples (6 mm in diameter) to be imaged at 2 μm resolution (e.g. figure 3), punched from 20 leaves selected from high Au samples (~10–50 ppb Au bulk average). Some Au particles were observed to be associated with Ca oxalate crystals (figure 3b), which parallels the associations seen between Ca and Au in calcrete in arid areas, and suggests that Au and Ca also share a common biological pathway in these systems [29].

8. Conclusions

The real-time event-by-event processing approach used in the Maia system relaxes constraints on the selection of dwell time per pixel. Image sizes and pixel dwell can be chosen to suit the needs of the application and to provide adequate counting statistics, with pixel size set generally by the beam size. Obtaining adequate counting statistics, especially for trace elements, is aided by the large collection solid-angle of Maia and the pulse handling and real-time processing modules, which can handle event rates beyond 20 M/s. Alternatively, large solid-angle and small dwell times can be used to minimize dose per pixel and help avoid radiation damage induced chemical changes to samples. Combined with the reliability and stability of the XFM beamline, the result is a flexible microanalytical tool for large area, high definition mapping at ~2 μm spatial resolution to around 10^8 pixels, and for 3D imaging modalities, such as XANES chemical state imaging and 3D fluorescence tomography. Early results at the P06 beamline show that similar flexibility and performance are being provided by the new Maia 384B detector, which exhibits improved energy resolution, yielding high definition elemental maps at ~0.5 μm spatial resolution.

In user applications, Maia is proving especially useful for large scale, high definition XFM imaging and locating rare minute metal particles in geological and biological samples. The large detector array, with a wide range of X-ray take-off angles, also provides sensitivity to the depth of features within a sample, which has been utilized to provide an imaging depth contrast. Maia large area mapping was exploited to capture the complexity of chemical and physical processes at play in the Sunrise Dam deposit and their influence on the mobility and deportment of gold in the ore. Rare gold precipitates were observed in the leaves of Eucalypts growing above an undisturbed ore body concealed beneath ~30 m of cover. The results provide a sound basis for the application of gold in selected vegetation species as a technique for the exploration for gold in concealed ore deposits.

Acknowledgements
This research was undertaken on the X-ray fluorescence microscopy beamline at the Australian Synchrotron, Victoria, Australia, and the Hard X-ray Micro/Nano-Probe beamline P06, Petra III, DESY, Germany. The authors wish to thank Stefan Vogt for his help acquiring SDD standards spectra.

References

[1] Siddons D P, Kirkham R, Ryan C G, De Geronimo G, Dragone A, Kuczewski A J, Li Z Y, Carini G, Beuttenmuller R, Elliott D *et al.* 2013 Maia X-ray Microprobe Detector Array System *these proceedings*
[2] Paterson D, de Jonge M D, Howard D L, Lewis W, McKinlay J, Starritt A, Kusel M, Ryan C G, Kirkham R, Moorhead G, Siddons D P 2011 *AIP Conf. Proc.* **1365** 219-222
[3] Schroer C G, Kurapova O, Patommel, Boye P, Feldkamp J, Lengeler B, Schwab A, Stephan S, Falkenberg G, Wellenreuther G and Reimers N 2010 *Nucl. Instr. Meth. A* **616** 93
[4] Kirkham R, Dunn P A, Kucziewski A, Siddons D P, Dodanwela R, Moorhead G, Ryan C G, De Geronimo G, Beuttenmuller R, Pinelli D *et al.* 2010 *AIP Conf. Proc.* **1234** 240-243
[5] Ryan C G, Kirkham R, Hough R M, Moorhead G, Siddons D P, de Jonge M D, Paterson D J, De Geronimo G, Howard D L and Cleverley J S 2010 *Nucl. Instr. Meth A* **619** 37-43
[6] De Geronimo G, O'Connor P, Beuttenmuller R H, Li Z, Kuczewski A J and Siddons D P 2003 *IEEE Transactions on Nuclear Science* **50** 885–891
[7] Dragone A, De Geronimo G, Fried J, Kandasamy A, O'Connor P and Vernon E 2005 *IEEE Nuclear Science Symposium Conference Record* **2** 914–918
[8] Ryan C G, Etschmann B E, Vogt S, Maser J, Harland C L, van Achterbergh E and Legnini D 2005 *Nucl. Instr. Meth. B* **231** 183-188

[9] Solé V A, Papillon E, Cotte M, Walter Ph, Susini J 2007 *Spectrochimica Acta B* **62** 63-68
[10] Bevington P R 1969 *Data reduction and error analysis for the physical sciences* (McGraw-Hill, New York)
[11] Ryan C G and Jamieson D N 1993 *Nucl. Instr. Meth. B* **77** 203-214
[12] Ryan C G 2000 *International Journal of Imaging Systems and Technology* (special issue on "Advances in Quantitative Image Analysis") **11** 219-230
[13] Ryan C G, Cousens D R, Sie S H, Griffin W L, Suter G F and Clayton E 1990 *Nucl. Instr. Meth. B* **47**, 55-71
[14] Ryan C G, Siddons D P, Kirkham R, Dunn P A, Kuczewski A, Moorhead G, De Geronimo G, Paterson D J, de Jonge M D, Hough R M *et al.* 2010 *X-Ray Optics and Microanalysis, AIP Conference Proc.* **1221** 9-17
[15] Ryan C G, Siddons D P, Kirkham R, Li Z Y, de Jonge M D, Paterson D, Cleverley J S, Kuczewski A, Dunn P A, Jensen M *et al.*2013 The Maia detector array and X-ray fluorescence imaging system: Locating rare precious metal phases in complex samples (SPIE conference on X-ray Nanoimaging, San Diego Aug 2013) *AIP Conf. Proc.* in press.
[16] Alfeld M 2013 The processing of large, heterogeneous data sets acquired for XRF imaging *ICXOM conference* presentation and abstract
[17] Ryan C G, Jamieson D N, Churms C L and Pilcher J V 1995 *Nucl. Instr. Meth. B* **104** 157-165
[18] Howard D L, de Jonge M D, Lau D, Hay D, Varcoe-Cocks M, Ryan C G, Kirkham R, Moorhead G, Paterson D and Thurrowgood D 2012 *Analytical Chemistry* **84** 3278-3286
[19] Zhang J Z, Bryce N S, Siegele R, Carter E A, Paterson D, de Jonge M D, Howard D L, Ryan C G, Hambley T W 2012 *Integrated Biology* **4** 1072-1080
[20] Kopittke P M, Menzies N W, de Jonge M D, McKenna B A, Donner E, Webb R I, Paterson D, Ryan C G, Glover C J and Lombi E 2011 *Plant Physiology* **156** 663-673
[21] Lombi E, de Jonge M D, Donner E, Kopittke P M, Howard D L, Kirkham R, Ryan C G and Paterson D 2011 *PLoS One* (Public Library of Science) **6(6)** e20626
[22] Carey A-M, Lombi E. Donner E. de Jonge M D, Punshon T, Jackson B P, Guerinot M L, Price A H, Meharg A A 2012 *Analytical and Bioanalytical Chemistry* **402** 3275-3286
[23] Lintern M, Hough R and Ryan C 2012 *J. Geochem. Explor.* **112** 189-205
[24] Barnes S, Godel B, Locmelis M, Fiorentini,M, Ryan C 2011 *Aust. J. Earth Sciences* **58** 691-709
[25] Etschmann B E, Ryan C G, Brugger J, Kirkham R, Hough R M, Moorhead G F, Siddons D P, De Geronimo G, Kuczewski A, Dunn P A, *et al.* 2010 *American Mineralogist* **95** 884-887
[26] Brugger J, Pring A, Reith F, Ryan C G, Etschmann B, Liu W, O'Neill B, Ngothai Y 2010 *Radiation Physics and Chemistry* **79** 151-161
[27] Hough R M, Noble R R P, Butt C R M, Ryan C G, Lintern M J, Reddy S M, Paterson D, de Jonge M D 2010 *Smart Science For Exploration And Mining* **1-2** 716-718
[28] Cleverley J S, Hough R, Ryan C, Maas R, Blake K, Nugus M 2010 *Smart Science For Exploration And Mining* **1-2** 213-215
[29] Lintern M, Anand R, Ryan C, Paterson D 2013 *Nature Communications* **4** 2614 (http://dx.doi.org/10.1038/ncomms3614)
[30] Barnes S J, Godel B, Ryan C G 2011 *Mineralogical Magazine* **75** 489
[31] Dyl K A, Cleverley J S, Bland P A, Ryan C G and Hough R M 2013 Trace element mapping of carbonaceous chondrites Allende (CV3-Oxidized) and Vigarano (CV3-Reduced) at 2-micron resolution over centimetre-scales *Geochimica et Cosmochimica Act* submitted
[32] Padovan A, Munksgaard N, Alvarez B, McGuinness K, Parry D, Gibb K 2012 *Hydrobiologia* **687** 275-288
[33] Donner E, Ryan C G, Howard D L, Zarcinas B, Scheckel K G, McGrath S P, de Jonge M D, Paterson D, Naidu R, Lombi E 2012 *Environmental Pollution* **166** 57-64
[34] Myers D E, Ryan C G, Kirkham R, Paterson D, Moorhead G, Dunn P A, Siddons D P, de Jonge M D, Howard D L, De Geronimo G *et al.* 2010 *Injury* **41** (supplement 1) S30-S31
[35] Lombi E, Smith E, Hansen T H, Paterson D, de Jonge M D, Howard D L, Persson D P, Husted

S, Ryan C and Schjoerring J K 2011 *J Experimental Botany* **62** 273–282
[36] Lye J C, Hwang J E C, Paterson D, de Jonge M D, Howard D L, Burke R 2011 *PLoS One* **6** e26867-1 - e26867-8
[37] McColl G, James S A, Mayo S, Howard D L, Ryan C G, Kirkham R, Moorhead G F, Paterson D, de Jonge M D, Bush A I 2012 *PLoS ONE* **7** e32685-1 - e32685-4
[38] Kopittke P M, de Jonge M D, Menzies N W, Wang P, Donner E, McKenna B A, Paterson D, Howard D L and Lombi E 2012 *Plant Physiology* **159** 1149–1158
[39] Wang P , Menzies N W, Lombi E, McKenna B A, de Jonge M D, Donner E, Blamey F P C, Ryan C G, Paterson D J, Howard D L *et al.* 2013 *Science of the Total Environment* **463-464** 131–139
[40] Howell N, Lavers J, Uematsu S, Paterson D, Howard D L, Spiers K M, de Jonge M D, Garrett R, Banati R 2013 High resolution X-Ray fluorescence microscopic analysis of breast feathers from two species of Procellariiformes, the Flesh-footed Shearwater (Puffinus carneipes) and the Streaked Shearwater (Calonectris leucomelas) *Integrative Biology* submitted
[41] Lombi E, de Jonge M D, Donner E, Ryan C G, Paterson D 2011 *Anal. Bioanal. Chem.* **400** 1637-1644
[42] Macdonald D, Rougieux F, Mansoulie Y, Tan J, Paterson D, Howard D L, de Jonge M D and Ryan C G 2010 *Phys. Status Solidi A* **207** 1807-1810
[43] Monico L, Janssens K, Alfeld M, Vanmeert F, Falkenberg G, Ryan C G, Howard D L, Cotte M, Miliani C 2013 Van Gogh and alteration of chrome yellow pigments: new Cr- speciation studies using the Maia X-ray detector *ICXOM conference* presentation and abstract
[44] van Gysel M, Lemberge P, van Espen P 2003 *X-ray Spectrometry* **32** 139
[45] Spiers K M, pers. comm., manuscript in prep.
[46] GeoPIXE software suite (http://nmp.csiro.au/GeoPIXE.html)
[47] de Jonge M D and Vogt S 2010 *Current Opinion in Structural Biology* **20** 606–614
[48] de Jonge M D, pers. comm., manuscript in prep.
[49] Micromatter XRF calibration standards (MICROMATTER 4004 Wesbrook Mall, Vancouver, British Columbia, V6T 2A3, Canada) (http://www.micromatter.com)
[50] Ryan C G, Siddons D P, Kirkham R, Paterson D J, de Jonge M D *et al.* Maia X-ray fluorescence imaging system: Real-time methods, imaging performance and applications, in prep.
[51] Ryan C G, Clayton E, Griffin W L, Sie S H and Cousens D R, 1988, *Nucl. Instr. Meth. B* **34** 396-402
[52] Vogt S, Pers. Comms.
[53] Yardley B W D and Cleverley J S 2013 The role of metamorphic fluids in the formation of ore deposits, in *Ore Deposits in an Evolving Earth* eds. Jenkin G R T, Lusty P A J, Mcdonald I, Smith M P, Boyce A J and Wilkinson J J (Geological Society, London) *Special Publications* **393** (http://dx.doi.org/10.1144/SP393.5)
[54] Cleverley J S, Hough R M, Ryan C G, Borg S 2011 *Let's Talk Ore Deposits* **1-2** 572-574
[55] Halfpenny A, Fisher L, Hough R, Cleverley J 2013 *12th SGA Biennial Meeting Proc.* Mineral Deposit Research for a High-Tech World **3** 1255-1258
[56] Lintern M J 2007 *Geochem. Exploration Environ. Anal.* **7** 249–266

Wavefront preserving channel-cut optics for coherent x-ray scattering experiments at the P10 beamline at PETRAIII

A. V. Zozulya[1*], A. Shabalin[1], H. Schulte-Schrepping[1], J. Heuer[1], M. Spiwek[1], I. Sergeev[1], I. Besedin[1,2], I. A. Vartanyants[1,2], and M. Sprung[1]**

[1]*Deutsches Elektronen-Synchrotron DESY, Notkestraße 85, 22607 Hamburg, Germany*
[2]*National Research Nuclear University "MEPhI", 115409 Moscow, Russia*

[*]alexey.zozulya@desy.de; [**]michael.sprung@desy.de

Abstract. We report on the performance of cryogenically cooled monolithic Si(111) channel-cut monochromator optics installed at the coherence beamline P10 of the PETRA III synchrotron source. Our results show that a high quality channel-cut monochromator crystal preserves the beam wavefront and provides excellent beam stability, which offers considerable benefits for coherent x-ray scattering experiments.

1. Introduction

High heat load monochromators are key components of many hard x-ray undulator beamlines. Monochromator optics at third-generation synchrotron sources usually need to fulfill the following tasks: i) provide high monochromaticity, ii) separate the monochromatic beam from the white beam and *Bremsstrahlung* radiation and iii) protect downstream components from high heat load (total power up to 600 W) of the polychromatic undulator beam. To achieve these needs double crystal monochromator (DCM) optics are commonly used, where a pair of separate silicon crystals are aligned for the same Bragg reflection in parallel configuration. In the DCM design the second crystal can be translated parallel and perpendicular to its surface enabling a wide range of photon energies to be covered, and additionally keeping the exit beam offset constant. Another advantage is that separate crystal plates can be polished to a high degree of surface quality. The primary disadvantage of DCM design relates to the requirement for active cryogenic cooling of both monochromator crystals, in order to avoid thermal distortions caused by exposure to the high heat load of undulator radiation. Inevitably cryocooling induces *independent* angular vibrations of both crystals, which cause unwanted angular motions of the x-ray beam and thereby increasing the effective source size.

One solution to reduce such directional vibrations of the monochromatic x-ray beam is to employ a channel-cut crystal as the high heat load monochromator. Channel-cut optics are fabricated from a monolithic slab of silicon crystal by cutting a groove which separates two diffracting crystal surfaces [1-8]. This rigid design naturally aligns both diffracting crystal surfaces and prevents their *independent* angular motions, thus providing a directionally stable beam. Essentially, channel-cut optics transfer spatial vibrations into small variations in energy space. The major drawback of a channel-cut monochromator design is that the exit beam offset is a function of the Bragg angle for the selected photon energy of monochromatic x-ray beam. Another restriction is caused by the fixed length of the second crystal which imposes the upper limit to the accessible range of photon energies.

The homogeneity or smoothness of the beam wavefront is another characteristic which is influenced by the quality of x-ray optics. The wavefront homogeneity is of special importance for coherent x-ray scattering experiments using X-ray Photon Correlation Spectroscopy (XPCS) [9-11] and Coherent X-ray Diffraction Imaging (CXDI) [12-14] techniques. In XPCS and CXDI experiments wavefront distortions can lead to a deterioration of speckle contrast thus hindering data interpretation. The quality of the wavefront is directly coupled to the surface roughness of the crystal optics. In the case of a channel-cut crystal geometry the surfaces are not easily accessible, and the required quality of surface treatment is very difficult to achieve.

In this work we report on the performance of a high quality polished channel-cut monochromator installed at the coherence beamline P10 of the PETRA III synchrotron facility at DESY, Hamburg [15].

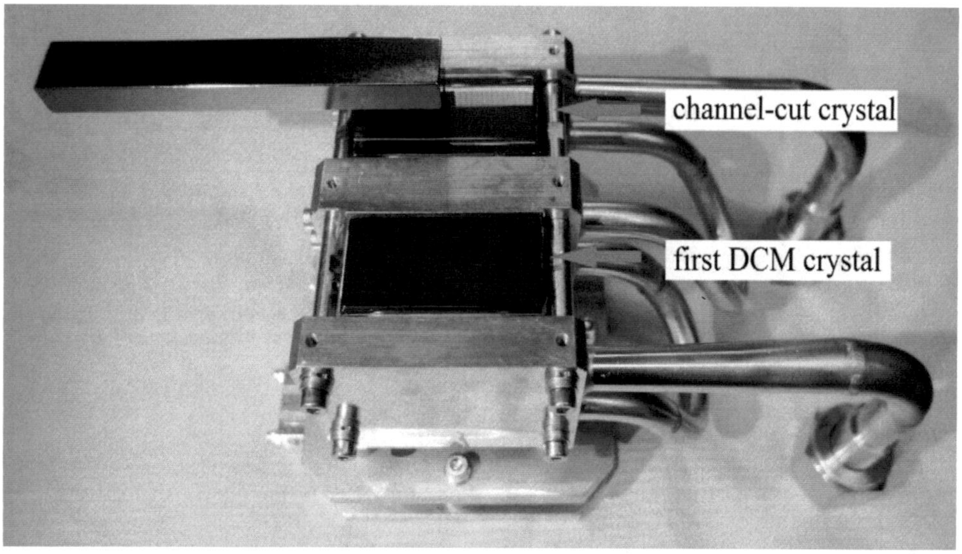

Figure 1. Channel-cut crystal and first crystal of standard double crystal monochromator mounted on the common cryogenically cooled copper block.

2. Experiment

The coherence beamline P10 is located at Sector 7 of the new low-emittance storage ring PETRA III having 2.3 km circumference [16]. The source is a 5 m length U29 undulator with 168 poles and 29 mm period, which delivers maximum power of 7.5 kW at 100 mA ring current. The insertion device is located in a low-beta section of the storage ring and provides an x-ray source with a vertical size of 14 μm and a horizontal size of 85 μm (full-width at half maximum values). The undulator beam subsequently passes through a high heat load Si(111) monochromator (vertically reflecting double crystal or channel-cut optics) located at 38.5 m distance from the source. Using double crystal monochromator enables the photon energy range between 4 and 25 keV. For rejecting the higher-order harmonics, a pair of horizontally reflecting flat silicon mirrors with curvature radius R>100 km is installed after the monochromator at the optics hutch of the P10 beamline.

An important design feature of PETRA III monochromators is the cryogenically cooled copper block with two slots for installing crystal optics. The cryogenic cooling is performed by a closed loop liquid nitrogen cooler, which supplies liquid nitrogen at a controlled and stable pressure to the monochromator crystal optics. Figure 1 shows a photograph of the copper block carrying the first crystal of DCM (lower slot) and the channel-cut crystal (upper slot). The crystal support block is mounted inside the ultra-high vacuum (UHV) chamber on the rotation axis of the monochromator. The

Journal of Physics: Conference Series **499** (2014) 012003 doi:10.1088/1742-6596/499/1/012003

rotation axis is equipped with a closed loop servomotor and enables high precision angular positioning with an accuracy of 2×10^{-5} degrees.

The channel-cut crystal was prepared and fabricated for symmetric Si(111) Bragg reflection. The DESY-FS crystal optics laboratory provided an aligned blank Si crystal, the cutting and polishing has been done by Holm Silicium-Bearbeitung GmbH. The channel-cut crystal has a gap between diffracting surfaces of 11.4 mm which results in 22.2 mm vertical offset between polychromatic and monochromatic beams at 8 keV photon energy. The length of the second surface of the channel-cut is 100 mm which sets an upper limit of 16 keV for the photon energy range of the monochromatic beam (lower limit of 5 keV is constrained by the beamstop collimator design). The presented P10 monochromator design allows to either use the standard DCM or channel-cut optics. The changeover from the DCM to the channel-cut optics is performed by 40 mm horizontal translation of the monochromator UHV chamber.

3. Results and discussion

2.1. Beam profile measurement

Before installing the channel-cut optics at the P10 monochromator UHV chamber, the profiles of the x-ray beam after the DCM were measured without and with the channel-cut optics in the beam. The DCM was selecting the photon energy of 7.9 keV. The channel-cut crystal was positioned in the first experimental hutch of P10 beamline at a distance of 71.5 m from the undulator source (33 m downstream from the DCM). The channel-cut crystal was aligned to diffract the beam in vertical direction in dispersive configuration relative to the DCM optics. The size of incident monochromatic beam at the channel-cut position was 2×0.63 mm^2 (horizontal×vertical) as defined by frontend slits. The beam profiles were registered using x-ray CCD camera coupled to a scintillator screen. The x-ray camera provided a resolution of about 8 micron and was placed at the detector table at a distance of 93.5 m from the source. The beam profile images obtained for the DCM and the combination of DCM followed by channel-cut crystal are shown in Fig. 2. They are practically identical and reveal smooth intensity distribution defined by the source shape. Slight asymmetry is due to frontend power slits geometry. The beam profile image obtained with the channel-cut crystal reproduces the profile of DCM beam and reveals only several vertical lines originating from surface treatment. Single dot artifacts visible in Fig. 2 originate from sparse dust particles adsorbed to intensity monitor foils.

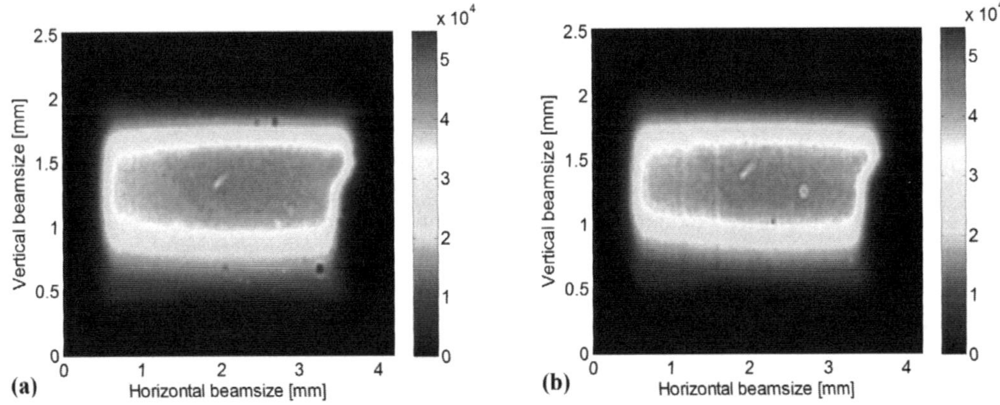

Figure 2. Beam profile images measured using DCM at 7.9 keV photon energy (a) without channel-cut crystal and (b) with channel-cut optics.

2.2. X-ray diffraction surface mapping

The bulk quality of the channel-cut crystal was evaluated by x-ray diffraction rocking curve measurements using standard Si(111) double crystal monochromator at two photon energies of 7.9 keV and 15 keV. The channel-cut crystal was positioned as described in previous paragraph and the diffracted intensity was registered by a silicon photodiode inserted in the beam path behind the channel-cut. Fig. 3 shows the rocking curves measured at room temperature for the symmetric Si(111) reflection of the channel-cut crystal in vertical diffraction plane. Experimental curves are in a good agreement with the curves calculated using dynamical diffraction theory for the given reflections in double crystal monochromator and channel-cut crystal. The energy dispersion was taken into account by integrating of reflection profiles over the source bandwidth. From measurements of rocking curves at 7.9 keV and 15 keV photon energies we obtained full width at half maximum (FWHM) values of 7.2 and 4.87 arcsec. These FWHMs are very close to the theoretical Darwin widths of 7.07 and 3.6 arcsec. for Si(111) Bragg reflection at the considered x-ray photon energies [17]. Experimental rocking curves were normalized to the incident intensities and yielded reflectivity values of 73% and 70% at 7.9 keV and 15 keV photon energies, respectively.

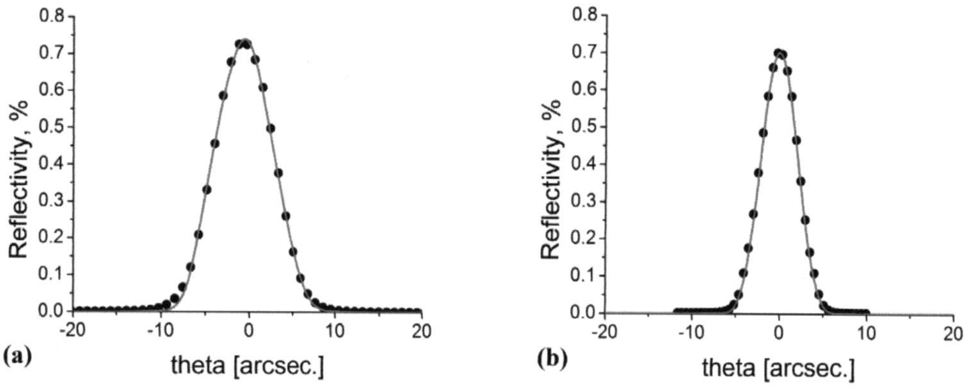

Figure 3. Rocking curves of Si(111) reflection from the channel-cut crystal measured using Si(111) DCM at two photon energies, (a) 7.9 keV and (b) 15 keV. Experimental points are shown by dots and calculated curves are plotted by solid lines.

The quality of surface treatment of the channel-cut crystal was evaluated by surface x-ray diffraction mapping. For surface mapping measurement we used a high resolution x-ray camera PCO Edge equipped with magnifying optics and scintillation screen, which provided about 1 μm resolution. The PCO Edge camera was positioned at the distance of 93.5 m from the source, thus having 22 m distance downstream from the channel-cut. The x-ray camera was used to collect projection images of the reflected beam while raster scanning of the channel-cut crystal over the incident monochromatic beam. At each position of the raster mesh the channel-cut crystal was realigned to the maximum of a rocking curve. We collected raster meshes with 0.5 mm step in vertical and horizontal directions which resulted in the full accessible crystal surface areas of 9.5×47 mm^2 (horizontal×vertical) and 9.5×40 mm^2 at the photon energies of 7.9 keV and 15 keV, respectively. The obtained surface maps are shown in Fig. 4. Every single-view image in these maps was divided by an averaged beam profile and normalized to the total incident intensity. The single-view images were stitched together to produce the full-field surface maps. The resulting maps confirm high quality of both crystal surfaces of the channel-cut monochromator. Large central areas of the crystal surfaces are artifact free which guarantees beam wavefront preservation after passing through the channel-cut optics. One can observe only a few mainly longitudinal striations, apparently due to polishing treatment. After these measurements the channel-cut crystal was installed in its working position at the UHV chamber of P10 monochromator.

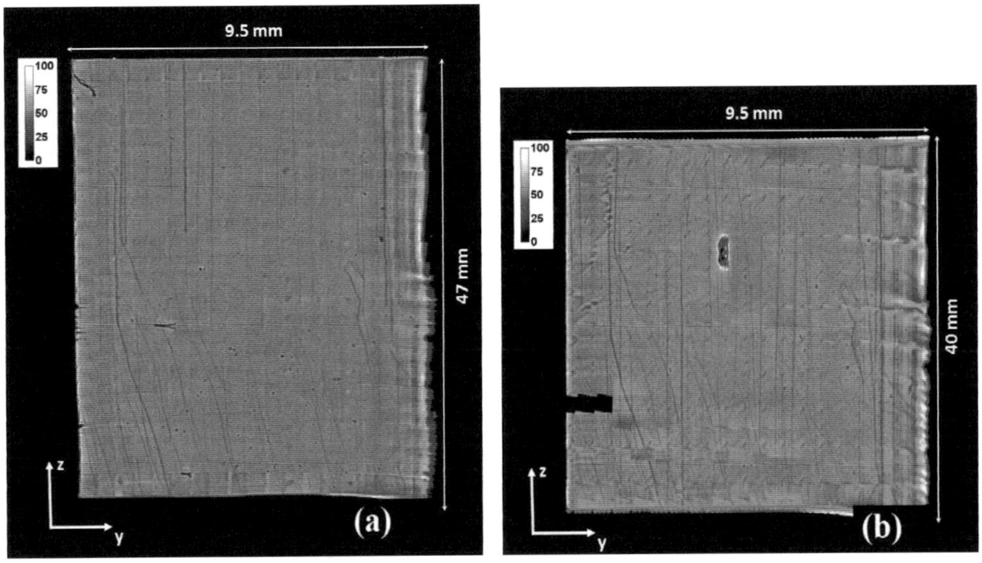

Figure 4. Surface x-ray diffraction maps of the channel-cut crystal measured at two photon energies, (a) 7.9 keV and (b) 15 keV.

2.3. Beam stability measurements

Measurements of beam stability after the channel-cut monochromator were performed using transfocator optics based on compound refractive lenses. The transfocator was installed at the second experimental hutch of the P10 beamline at a distance of 86 m from the source [18]. Using the transfocator the monochromatic beam was focused onto the blades of motorized slit system located at 1.3 m distance downstream from the focusing optics. Vertical and horizontal slit blades were used for knife-edge scanning and cutting the focused beam to 50% level of incident intensity for stability measurements. The transmitted intensity was monitored using a storage oscilloscope. The time dependences of oscilloscope signal overlaid with knife-edge scans measured in vertical and horizontal directions are shown in Fig. 5. Each knife edge measurement was repeated three times and the resulting profiles are displayed in Fig. 5 by red lines. Focused beam size FWHM values of 1.3 μm in vertical and 2.5 μm in horizontal direction were determined from the widths of derivative curves. Intensity signal deviations were scaled to the slope of the knife edge profile to determine the amplitude (r.m.s value) of beam deviations as depicted by blue arrows and horizontal cyan lines in Fig. 5. The time dependence measurements performed for the channel-cut monochromator demonstrate high stability of the monochromatic beam. Spatial deviations of 50 nm (r.m.s. value) were observed in vertical direction and 70 nm in horizontal direction, which is equivalent to angular beam deviations of 45 nrad and 62 nrad, correspondingly. In comparison with the standard Si(111) DCM, the beam stability using Si(111) channel-cut monochromator has improved by more than five times in vertical direction and about 50% in horizontal direction. Measured efficiency of channel-cut monochromator at 7 keV photon energy is about 80% as compared to DCM conditions.

4. Summary

We describe the performance of cryogenically cooled channel-cut monochromator optics implemented at the coherence beamline P10 of the PETRA III synchrotron facility. A new monolithic channel-cut crystal has been fabricated and subsequently subject to fine surface treatment. From rocking curve measurements at room temperature we have obtained reflectivity values above 70% and the widths of rocking curves which are very close to theoretical values of Bragg reflection in perfect Si crystal. This result confirms high quality of the bulk of channel-cut crystal optics. Beam profile and surface x-ray diffraction mapping measurements performed at 7.9 keV and 15 keV photon energies indicate high surface quality of both diffracting surfaces of the channel-cut crystal. No wavefront deterioration was observed. Rigid design of monolithic channel-cut monochromator crystal results in excellent beam stability of less than 50 nrad measured in vertical plane under high heat load conditions. The performance of channel-cut monochromator optics offers considerable benefits for coherent x-ray scattering experiments at the P10 beamline.

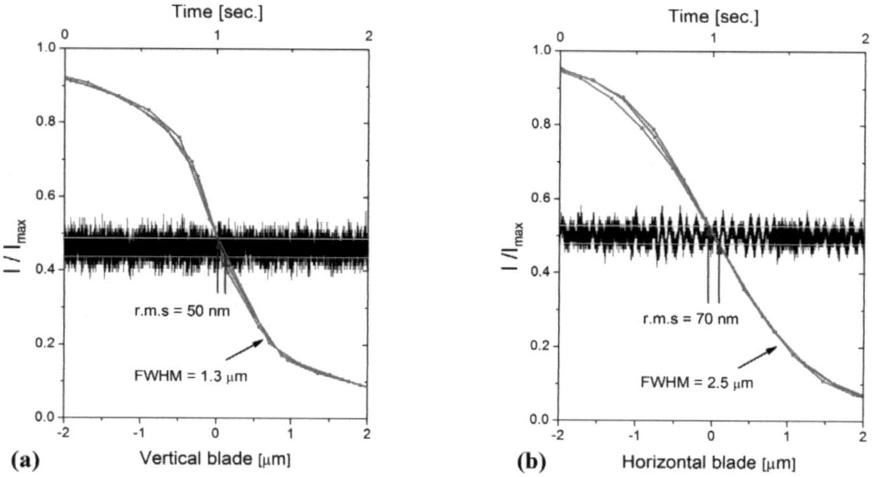

Figure 5. Time dependences of oscilloscope signal I/I_{max} (black lines) monitoring the intensity of the beam focused onto slit blades. I/I_{max} signal is overlaid with knife-edge scans in (a) vertical and (b) horizontal directions (red lines). Amplitude of beam deviations is depicted by blue arrows and cyan horizontal lines.

Acknowledgments
Authors would like to thank F. Westermeier, S. Smoljanin and M. Lohmann for technical assistance with installing PCO Edge x-ray camera. We thank D. Novikov for careful reading of the manuscript and valuable remarks. Part of this work was supported by BMBF Proposal 05K10CHG "Coherent Diffraction Imaging and Scattering of Ultrashort Coherent Pulses with Matter" in the framework of the German-Russian collaboration "Development and Use of Accelerator-Based Photon Sources" and the Virtual Institute VH-VI-403 of the Helmholtz Association.

References

[1] Bonse U and Hart M 1965 *Appl. Phys. Lett.* 7 238
[2] Bonse U and Hart M 1966 *Z. Phys.* **189** 151
[3] Beaumont J H and Hart M 1974 *J. Phys. E: Sci. Instrum.* 7 823
[4] Berman L E, Durbin S M and Batterman B W 1985 *Nucl. Instr. and Meth.* A **241** 295
[5] Quintana J P, Hart M, Bilderback D, Henderson C, Richter D, Setterston T, White J, Hauserman D, Krumrey M and Schulte-Schrepping H 1995 *J. Synchrotron Rad.* **2** 1
[6] Berman L E, Siddons D P, Montanez P A, Lenhard A and Yin Z 2002 *Rev. Sci. Instrum.***73**,

1481
[7] Ishikawa T, Tamasaku K, Yabashi M 2005 *Nucl. Instr. and Meth.* A **547** 42
[8] Shu D, Narayanan S, Sandy A, Sprung M, Preissner C and Sullivan J 2007 *Proc. SPIE* **6665**, 66650O
[9] Grübel G and Zontone F 2004 *J. of Alloys and Compounds* **362**, 3
[10] Gutt C, Sprung M, Fendt R, Madsen A, Sinha S K and Tolan M 2007 *Phys. Rev. Lett.* **99**, 096104
[11] Pierce M S, Chang K C, Hennessy D, Komanicky V, Sprung M, Sandy A and You H 2009 *Phys. Rev. Lett.* **103** 165501
[12] Pfeifer M A, Williams G J, Vartanyants I A, Harder R and Robinson I K 2006 *Nature* **442** 63
[13] Vartanyants I A and Robinson I K 2003 *J. Synchrotron Radiat.* **10**, 409
[14] Gulden J, Yefanov O M, Mancuso A P, Dronyak R, Singer A, Bernátová V, Burkhardt A, Polozhentsev O, Soldatov A, Sprung M and Vartanyants I A 2012 *Optics Express* **20** 4039
[15] http://photon–science.desy.de/facilities/petra_iii/beamlines/p10_coherence_applications/index_eng.html
[16] Franz H, Leupold O, Röhlsberger R, Roth S V, Seeck O H, Spengler J, Strempfer J, Tischer M, Viefhaus J, Weckert E and Wroblewski T 2006 *Synchrotron Radiat. News* **19** 25
[17] http://sergey.gmca.aps.anl.gov/x0h.html
[18] Zozulya A V, Bondarenko S, Schavkan A, Westermeier F, Grübel G and Sprung M 2012 *Optics Express* **20** 18967

Journal of Physics: Conference Series **499** (2014) 012004 doi:10.1088/1742-6596/499/1/012004

Focusing XFEL SASE pulses by rotationally parabolic refractive x-ray lenses

Frank Seiboth[1], Andreas Schropp[1,2], Robert Hoppe[1], Vivienne Meier[1], Jens Patommel[1], Hae Ja Lee[2], Bob Nagler[2], Eric C. Galtier[2], Brice Arnold[2], Ulf Zastrau[2], Jerome B. Hastings[2], Daniel Nilsson[3], Fredrik Uhlén[3], Ulrich Vogt[3], Hans M. Hertz[3], Christian G. Schroer[1]

[1]Institute of Structural Physics, Technische Universität Dresden, D-01062 Dresden, Germany
[2]Linac Coherent Light Source, SLAC National Accelerator Laboratory, 2575 Sand Hill Road, Menlo Park, CA 94025, USA
[3]Biomedical & X-Ray Physics, KTH/Royal Institute of Technology, KTH-AlbaNova, SE-106 91, Stockholm, Sweden

E-mail: schroer@xray-lens.de

Abstract. Using rotationally parabolic refractive x-ray lenses made of beryllium, we focus hard x-ray free-electron laser pulses of the Linac Coherent Light Source (LCLS) down to a spot size in the 100 nm range. We demonstrated efficient nanofocusing and characterized the nanofocused wave field by ptychographic imaging [A. Schropp, et al., Sci. Rep. **3**, 1633 (2013)] in the case of monochromatic LCLS pulses produced by a crystal monochromator that decreases the LCLS bandwidth down to $\Delta E/E = 1.4 \cdot 10^{-4}$. The full spectrum of LCLS pulses generated by self-amplified spontaneous emission (SASE), however, fluctuates and has a typical bandwidth of a few per mille ($\Delta E/E \approx 2 \cdot 10^{-3}$). Due to the dispersion in the lens material, a polychromatic nanobeam generated by refractive x-ray lenses is affected by chromatic aberration. After reviewing the chromaticity of refractive x-ray lenses, we discuss the influence of increased bandwidth on the quality of a nanofocused SASE pulse.

1. Introduction

The ultra-short pulses of x-ray free-electron lasers open the way for the investigation of the dynamics of matter on atomic time and length scales. Focusing these pulses to 100 nanometers and below is crucial for many experiments, such as high-resolution and time-resolved imaging with single pulses, the study of non-linear optical phenomena, and the creation of matter in extreme conditions [1, 2, 3, 4, 5]. Several optics have been shown to be capable of focusing XFEL pulses to these small dimensions, such as Fresnel zone plates [6, 7], Kirkpatrick-Baez mirrors [8], and refractive x-ray lenses [9].

Today, refractive x-ray optics are widely spread at synchrotron radiation sources and are typically used in conjunction with a crystal monochromator. For monochromatic radiation, e. g., a relative bandwidth of $\Delta E/E = 1.4 \cdot 10^{-4}$ for Si (111) reflections, chromatic aberrations of refractive x-ray lenses are negligible. The typical bandwidth of an XFEL pulse generated by self-amplified spontaneous emission (SASE) is about one order of magnitude wider than that of a Si (111) crystal monochromator. In the case of the Linac Coherent Light Source (LCLS)

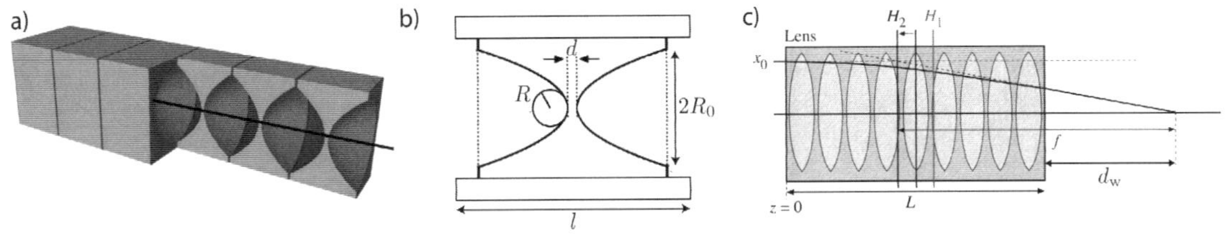

Figure 1. a) Rotationally parabolic refractive x-ray lens. The optic is made of a series of individual refractive x-ray lenses that are stacked behind each other. b) Parameters for individual refractive x-ray lenses. c) Thick refractive x-ray lens.

at the Stanford Linear Accelerator Center in Menlo Park, California, USA, the typical SASE bandwidth is about $\Delta E/E = 2 \cdot 10^{-3}$.

In this article we investigate the effect of this mild polychromaticity on the nanofocusing capabilities of parabolic refractive x-ray lenses using geometric arguments. Alternatively, an approach using longitudinal coherence could be followed, as demonstrated for diffractive optics [10]. For an ideal parabolic refractive lens and a more realistic optic that shows weak spherical aberrations, we model the intensity distribution around the nominal focus for monochromatic radiation, Gaussian spectral distributions of different width, and for a more realistic SASE spectrum. In a recent experiment, we focused monochromatic XFEL radiation to the 100 nanometer level by parabolic refractive x-ray lenses made of beryllium and characterized the nanobeam using the method of ptychography [9, 11]. Using the complex wave field reconstructed in this way, we calculate the chromatic aberrations of the optics and its consequences for nanofocusing.

2. Characterizing XFEL pulses by ptychography

We used ptychography to characterize the nanofocused x-ray pulses of the LCLS. A detailed account of this experiment can be found in [9]. The x-ray pulses were focused by parabolic refractive x-ray lenses made of beryllium [12, 13]. The compound lens was made of $N = 20$ single lenses with a radius of curvature $R = 50\,\mu\text{m}$. The aperture was defined by a pinhole to $300\,\mu\text{m}$ [cf. Fig. 1]. The expected diffraction limited focal spot size for this optic is 115 nm (full width half maximum, FWHM).

The experiments were carried out at the Matter in Extreme Conditions (MEC) endstation located 464 m downstream of the SASE undulator in the far experimental hall of LCLS. A Bartel type monochromator was used to filter the beam to a relative bandwidth of $\Delta E/E \approx 1.4 \times 10^{-4}$ at an energy of $E = 8.194\,\text{keV}$.

In order to characterize the nanofocused XFEL pulses, we recorded a ptychogram of a test sample. Ptychography is a scanning coherent diffraction imaging technique, in which a sample is scanned through a confined coherent x-ray beam in two dimensions perpendicular to the optical axis [14, 15]. At each position of the scan, a far-field diffraction pattern is recorded. The step size in between adjacent scan points is chosen such that the illuminated areas overlap. From these data, the complex transmission function of the object and the complex illuminating wave field in the plane of the sample can be reconstructed [16, 17, 18]. This method is ideal for quantitative x-ray imaging with highest spatial resolutions [19, 20, 21, 22] even in three dimensions [23, 24, 25]. Using a test object, the coherent beam can be fully characterized by ptychography [26, 27, 28, 29], giving access to the full caustic [30]. By now, the method is routinely applied to characterize x-ray nanofocusing optics at synchrotron radiation sources and the beam properties of hard x-ray scanning microscopes [31].

The test sample was placed into the XFEL beam about 0.5 mm downstream of the focal

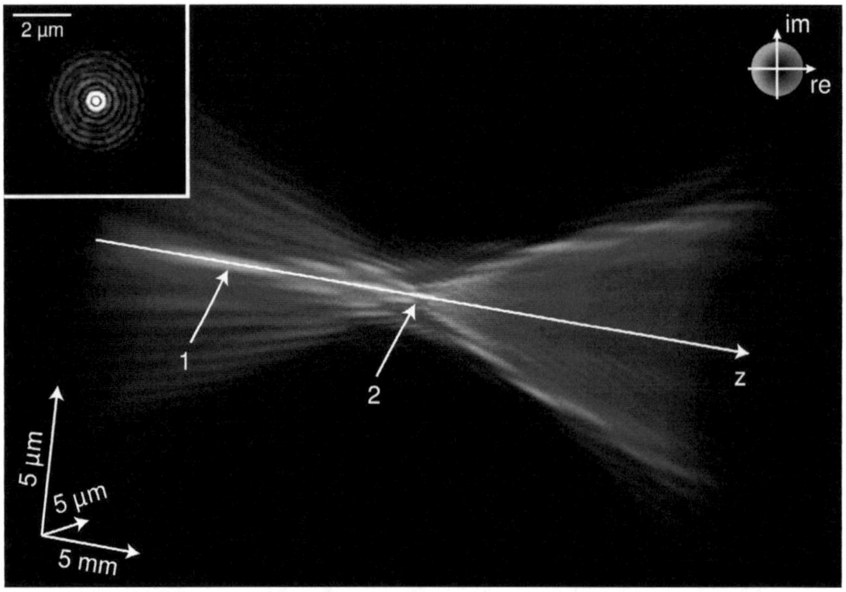

Figure 2. Reconstructed complex wave field propagated ±15 mm along the optical axis. Phase is encoded in color, amplitudes in hue. The caustic reveals a spherical aberration of the refractive x-ray optics, as the paraxial rays are focused to a shorter focal length (pointed to by arrow 1) whereas the peripheral rays coincide at a larger focal distance (pointed to by arrow 2). The inset shows the up-scaled wave field near position two.

plane and scanned on a two-dimensional grid perpendicular to the beam, covering an area of $2 \times 2\,\mu\text{m}^2$ with a step size of 100 nm. The test sample was nanostructured into a $1\,\mu$m-thick tungsten layer sputter deposited onto a $100\,\mu$m-thick CVD diamond substrate. The test patterns were arranged in an array of 40×40 identical star patterns, each approximately $2 \times 2\,\mu\text{m}^2$ in size with structures ranging from 50 nm to 200 nm. As monochromatized XFEL pulses strongly fluctuate in intensity, two diffraction patterns were recorded at each scan point with a Princeton Instruments PIXIS-XF 2048B x-ray camera located 4.14 m behind the focus.

For the ptychographic reconstruction, only diffraction patterns with intensities above a certain threshold were selected and normalized to the pulse intensity [9]. The ptychographic reconstruction yields both the complex transmission function of the object and a complex wave field. Since the ptychographic dataset is created by a large number of individual XFEL pulses, each slightly varying in shape and intensity, the complex illumination function represents an averaged illumination function produced by the nanofocusing optics. It yields detailed information about the nanofocused x-ray beam and is shown in the inset in Fig. 2. As the wave field is given in amplitude and phase, it can be used to reconstruct the full caustic near the nanofocus (Fig. 2) or to model near field images recorded in magnifying geometry [32].

In the following, we use this wave field to model the intensity distribution around the focus in the case of mild polychromaticity, i. e., for x-ray spectra with a bandwidth typical to SASE XFELs.

3. Chromaticity of x-ray lenses and limitations to focusing SASE pulses

The focal length of a compound refractive lens (CRL) is

$$f = \frac{1}{\omega \sin(\omega L)}, \quad \omega^2 = \frac{2\delta}{lR} = \beta\frac{2\delta}{R_0^2}, \tag{1}$$

where ω^2 is the refractive power per unit length. R is the radius of curvature of the parabolic surfaces of a single biconcave lens, l its overall length, and δ the index of refraction decrement of the lens material [cf. Fig. 1b) and eq. (2) below]. ω^2 can be expressed in terms of the geometric aperture radius R_0 by introducing a geometric factor $\beta := R_0^2/(lR) < 1$ [Fig. 1b)]. $L = Nl$ is the overall length of the compound lens made of N individual lenses [Fig. 1c)]. In this article, CRLs are treated as thick optics, unless L is small compared to the focal length f, in which case the lens can be treated as optically thin.

For a complex refractive index $n = 1 - \delta + i\beta$, δ is given by

$$\delta = \frac{n_a}{2\pi} r_0 \lambda^2 \left[f_0(K = 0) + f'(E) \right] \tag{2}$$

in the hard x-ray range. Here, n_a is the number density of the atoms in the lens material, r_0 the classical electron radius, λ the wavelength of the x rays, and $f_0 + f'$ the real part of the atomic form factor in forward direction. Away from absorption edges of the lens material, f' is nearly negligible and $f_0 = Z$ corresponds to the atomic number of the material. Therefore, δ scales inversely with the square of the energy ($\delta \propto E^{-2}$) for beryllium lenses in the hard x-ray range.

From eqs. (1) and (2) the relative change in focal distance f as a function of energy can be derived:

$$\frac{\Delta f}{f} = \underbrace{[1 + \omega L \cot(\omega L)]}_{=:\gamma} \frac{\Delta E}{E}. \tag{3}$$

For thin lenses, ωL is much smaller than one and $\Delta f / f = 2\Delta E/E$. For the thickest conceivable lens, i. e., when the focus is at the exit of the lens stack, $\gamma = 1$.

This chromaticity can be tolerated for nanofocusing, if the lateral beam size as a result of the chromaticity is smaller than the diffraction limited beam size, i. e.,

$$\Delta f \cdot NA \leq d_t = \alpha \frac{\lambda}{2NA}, \tag{4}$$

which is equivalent to require Δf to be smaller or equal to the depth of focus. In general, α depends on the aperture function of the optic and the resolution criterium. For refractive lenses with Gaussian aperture, d_t is the full width at half maximum of the Gaussian Airy disc and $\alpha = 2\sqrt{2 \ln 2}/\pi \approx 0.75$ [12]. NA is the numerical aperture given by

$$NA = \frac{D_{\text{eff}}}{2f}, \quad D_{\text{eff}} = 2\frac{\sqrt{2\delta}}{\omega} \left\{ \frac{\beta}{a_p} \left[1 - \exp(-a_p) \right] \right\}^{1/2}, \quad a_p := \beta \frac{\mu L}{2} \cdot \underbrace{\frac{1}{2}\left[1 + \frac{\sin(2\omega L)}{2\omega L} \right]}_{=:\epsilon}. \tag{5}$$

Here, μ is the linear attenuation coefficient of the lens material. D_{eff} is the so-called effective aperture. Using eq. (4), the maximal energy bandwidth that can be tolerated in a given imaging geometry is

$$\frac{\Delta E}{E} \leq \frac{2}{\alpha\gamma} \cdot \frac{d_t^2}{\lambda f} = \alpha \frac{2}{\gamma} \cdot \frac{\lambda f}{D_{\text{eff}}^2}. \tag{6}$$

The intensity in the nominal focal plane ($z = z_0$) is an incoherent sum over the intensities of the wave fields for the different spectral modes:

$$I_{z_0}(x, y) = \int \rho(E) |\psi(x, y, z_0, E)|^2 \, dE. \tag{7}$$

Here, $\rho(E)$ is the spectral density of the polychromatic beam and $\psi(x, y, z, E)$ is the spatial component of the wave field for a given energy E. The beam broadening in the nominal

Journal of Physics: Conference Series **499** (2014) 012004 doi:10.1088/1742-6596/499/1/012004

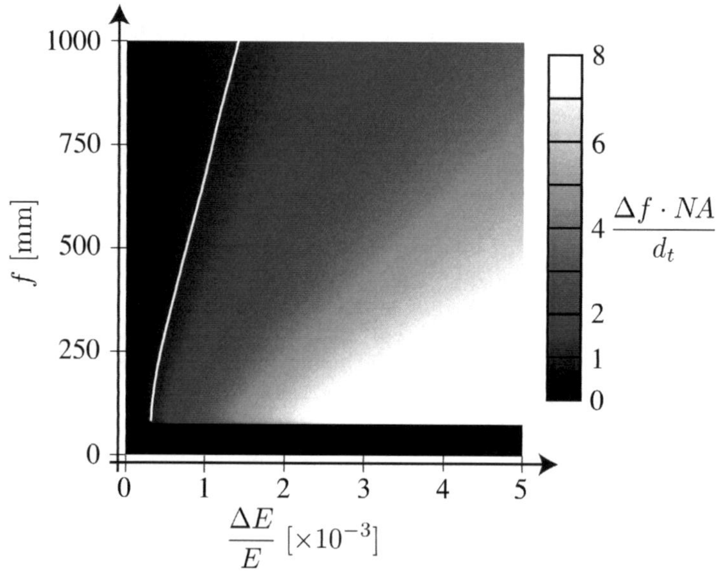

Figure 3. Broadening due to chromatic aberration [left-hand side of eq. (4)] relative to the diffraction limit d_t [right-hand side of eq. (4)] as a function of energy bandwidth and focal length. The white line delineates the border between diffraction limited (left) and chromaticity limited focusing (right). The black area in the lower part of the graph is masking a regime where the focus would lie within the lens.

focal plane can be estimated by considering the width b (FWHM) of the components of the polychromatic beam that are defocused by Δf:

$$b = \sqrt{d_t^2 + (\Delta f \cdot NA)^2} = \sqrt{d_t^2 + \left(\frac{D_{\text{eff}}}{2} \cdot \gamma \cdot \frac{\Delta E}{E}\right)^2}. \quad (8)$$

In the case of a nearly Gaussian aperture, i. e., $\exp(-a_p) \approx 0$, D_{eff} and d_t scale proportional to $\sqrt{f}$. In that regime, the relative influence of the chromaticity is independent of the focal length.

The refractive optics that were used in the nanofocusing experiments at LCLS are not in this Gaussian regime. The parameters are such that the Gaussian aperture is clipped by the geometric aperture. Therefore, the influence of the focal length on the chromaticity is not negligible in this case. Fig. 3 shows the fraction of the broadening due to chromatic aberration relative to the diffraction limit:

$$\frac{\Delta f \cdot NA}{d_t}.$$

If this quotient is smaller than one, the beam is dominated by the diffraction limit. If the quotient is larger than one, the nanofocus is dominated by chromatic effects. For the imaging parameters at the LCLS, i. e., $f = 250$ mm and $\Delta E/E \approx 2 \cdot 10^{-3}$, chromatic effects are already quite important, i. e., $\Delta f \cdot NA/d_t = 4.3$. In the following section, we investigate the impact of chromaticity for this particular case.

4. Focusing broadband x rays and SASE pulses with CRLs made of beryllium

To show the influences of broadband x rays on chromatic optics, we modeled the focusing properties for two sets of beryllium CRLs at the mean energy of $E = 8.194$ keV. The first one is

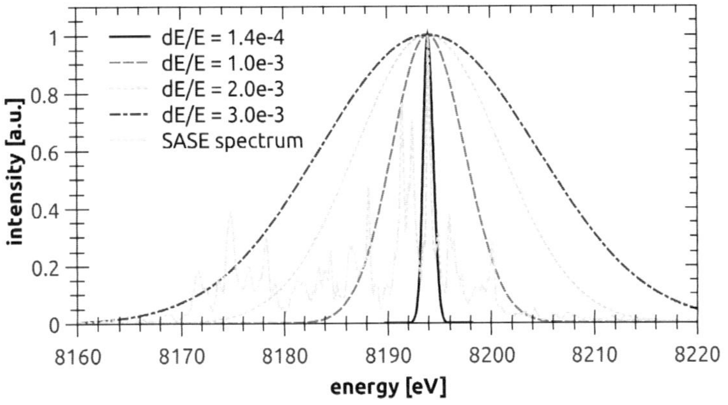

Figure 4. Spectral distributions used to calculate the influences of broadband XFEL beams. The Gaussian with a FWHM width of $\Delta E/E = 1.4 \cdot 10^{-4}$ models the monochromatic case. Three broader Gaussian distributions and a more realistic SASE spectrum are used to model the polychromatic beam from a SASE XFEL source.

an ideal lens (without spherical aberrations) with the same imaging parameters as the real lens that was used in the LCLS nanofocusing experiment ($N = 20, R = 50\,\mu\text{m}, 2R_0 = 300\,\mu\text{m}$). The focal spot size for this optic is expected to be 115 nm (FWHM) with a depth-of-focus of roughly 200 μm. Since the dominating effect of small energy variations is a change in focal distance [c. f. eq. (3)], the expected beam profile can be determined by overlapping a single monochromatic profile at different shifted positions along the optical axis. This assumes that the wave-front shape $\psi(x, y, z, E)$ [cf. eq. (7)] is not significantly changed for this mild polychromaticity, i. e., the focus remains similar in lateral distribution but shifts along the optical axis according to the chromatic aberration.

We calculate the intensity distribution in the nominal focal plane by the convolution of the 3-dimensional intensity distribution with the spectral distribution function mapped onto the optical axis by eq. (3). We used several spectral distribution functions and compared the focusing results. The spectral distribution functions are shown in Fig. 4.

The intensity distribution in the focal plane is shown for an ideal lens for the different energy spectra in Figure 5 a) and c). The focal spot size is already broadened at the base for $\Delta E/E = 1 \times 10^{-3}$. Reason for this is the perfect focusing characteristic of this lens that is free of spherical aberrations. The depth-of-focus is determined by the diffraction limit and is about 200 μm. Since the beam is smeared by approximately by half a millimeter along the optical axis for a polychromaticity of $\Delta E/E = 1 \times 10^{-3}$, the focal depth is clearly exceeded. Especially for the SASE spectrum one can see an even stronger broadening, since the spectral distribution is widely spread, with significant peaks well outside a Gaussian envelope (c. f. Fig. 4).

The situation is different for the optic that has a spherical aberration. To determine the intensity distribution, we made use of the measured wave field obtained during the ptychography experiment described in section 2. For this optic, the spherical aberrations lead to an elongation of the effective depth-of-focus as well as a small increase in focal spot size. Using this profile for the convolution with the exact same spectra as before, the line plots in Fig. 5 b) and d) are obtained. The focus has dominant side-lobes even for the nearly monochromatic case. When increasing the energy bandwidth, the main effect is a shift of the whole distribution to higher intensities. The shape is only changing by a margin. A more detailed look for selected spectra

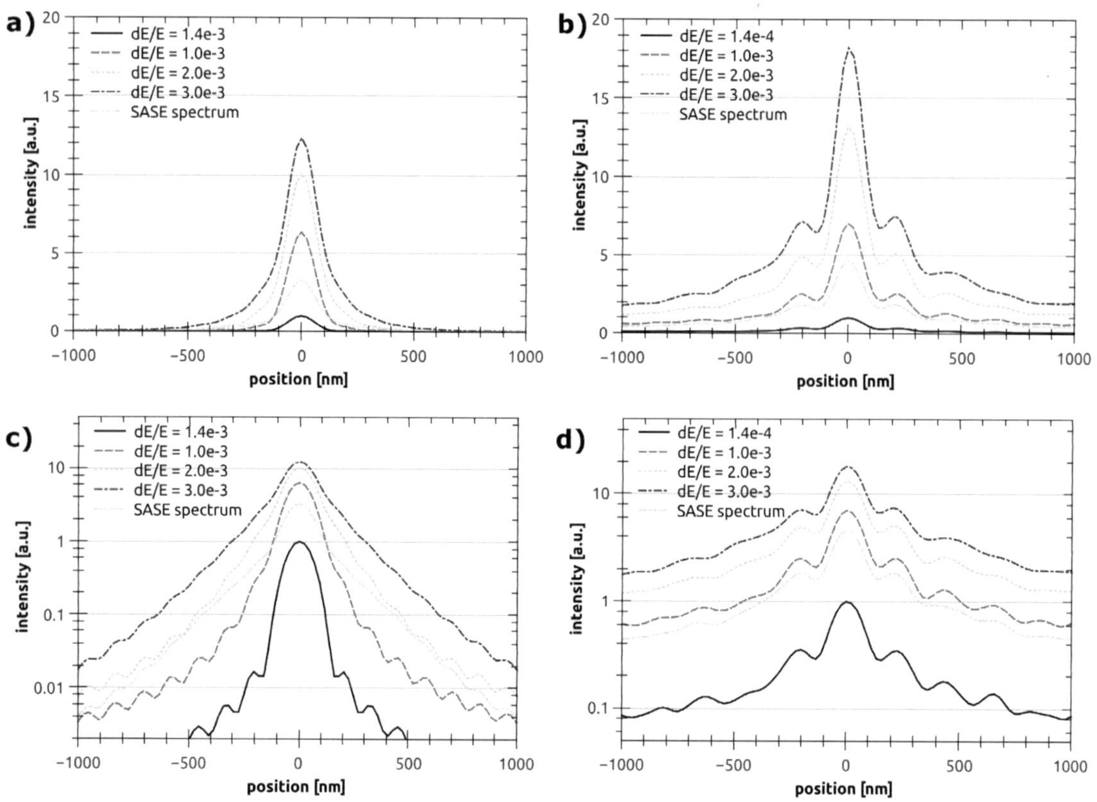

Figure 5. Integrated line profiles through the beam waist for an ideal parabolic optic on a linear a) and logarithmic scale c). Similar profiles for the lens with spherical aberration used during experiments at LCLS on a linear b) and logarithmic scale d).

is shown in Figure 6. The most dominant effect in the beam caustics on the left is a blurring, while the overall shape of the beam waist seems nearly unaffected. Also the sections through the beam shown in Fig. 6 on the right taken in the plane marked with a white line in the profiles (Fig. 6 left) show mostly a smearing of the side lobes, while the central peak remains almost unchanged in size.

The intensities in Fig. 5 are normalized to the peak intensity of the monochromatic beam. The relative gain in peak intensity for the polychromatic illumination is about one order of magnitude. The nanobeam generated by the ideal lens is more sensitive to chromatic aberration and thus the relative gain in peak intensity is slightly less than for the lens spherical aberration [compare Fig. 5a) and b)]. However, when comparing absolute peak intensities, the ideal optic is about a factor of five higher than the one with spherical aberration for the monochromatic case.

5. Conclusions

In general, the bandwidth of SASE pulses from the Linac Coherent Light Source is too large to be able to neglect chromatic effects when nanofocusing XFEL pulses with compound refractive lenses. While the full-width-half-maxium size of the beam is only slightly affected, the chromatic aberration results in a significant increase in the intensity in the shoulders and the side maxima of the beam. If the optics have spherical aberrations, the chromatic effects may be obscured by these aberrations and become dominant only for larger bandwidths. This is the case for the

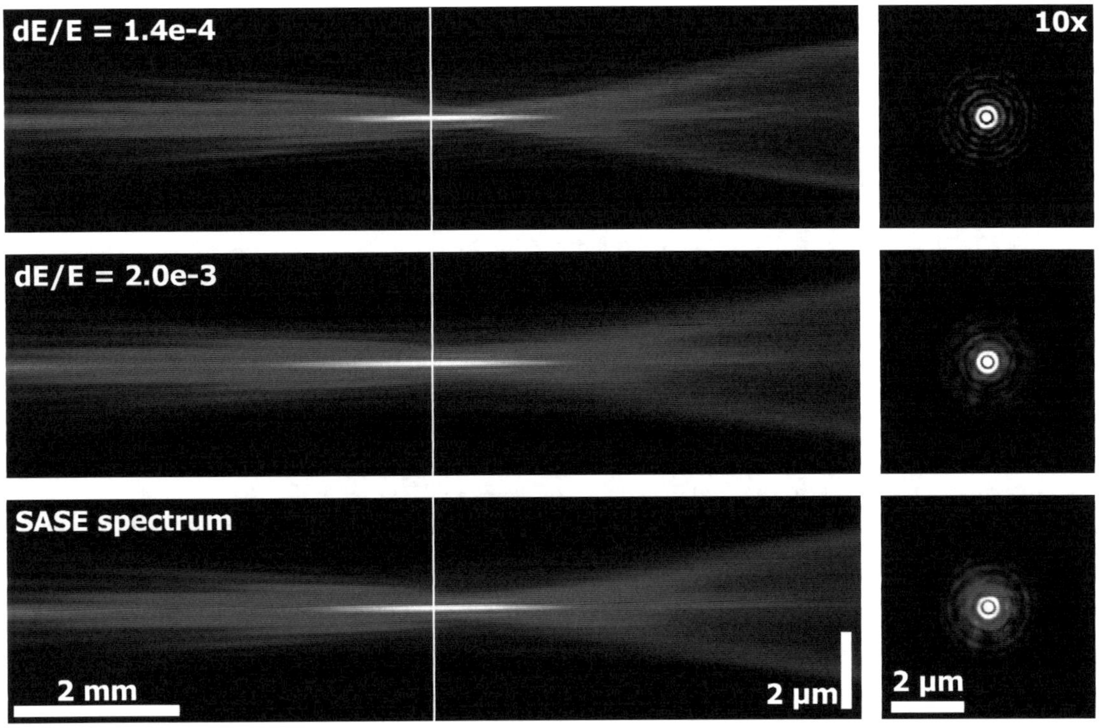

Figure 6. Calculated beam intensity along the caustic (left) and intensity in the focal plane marked by the white line (upscaled to show low intensity features) for the aberrated Be CRLs at different energy bandwidths (right).

parabolic refractive lenses used in our experiment at LCLS.

Acknowledgments

Portions of this research were carried out at the Linac Coherent Light Source (LCLS) at the SLAC National Accelerator Laboratory. LCLS is funded by the U.S. Department of Energy's Office of Basic Energy Sciences. The MEC instrument was supported by U.S. Department of Energy, Office of Fusion Energy Sciences. This work was funded by Volkswagen Foundation, the DFG under grant SCHR 1137/1-1, the German Ministry of Education and Research (BMBF) under grant number 05K13OD2, the Swedish Research Council and the Göran Gustafsson Foundation. We would also like to thank the MEC team at SLAC, collaborating institutions, and Bruno Lengeler for fruitful discussions and providing the CRL optics.

References

[1] Koopmann R, Cupelli K, Redecke L, Nass K, Deponte D P, White T A, Stellato F, Rehders D, Liang M, Andreasson J, Aquila A, Bajt S, Barthelmess M, Barty A, Bogan M J, Bostedt C, Boutet S, Bozek J D, Caleman C, Coppola N, Davidsson J, Doak R B, Ekeberg T, Epp S W, Erk B, Fleckenstein H, Foucar L, Graafsma H, Gumprecht L, Hajdu J, Hampton C Y, Hartmann A, Hartmann R, Hauser G, Hirsemann H, Holl P, Hunter M S, Kassemeyer S, Kirian R A, Lomb L, Maia F R N C, Kimmel N, Martin A V, Messerschmidt M, Reich C, Rolles D, Rudek B, Rudenko A, Schlichting I, Schulz J, Shoeman M M S R L, Sierra R G, Soltau H, Stern S, Strueder L, Timneanu N, Ullrich J, Wang X, Weidenspointner G, Weierstall U, Williams G J, Wunderer C B, Fromme P, Spence J C H, Stehle T, Chapman H N, Betzel C and Duszenko M 2012 *Nature Methods* **9** 259
[2] Johansson L C, Arnlund D, White T A, Katona G, Deponte D P, Weierstall U, Doak R B, Shoeman R L, Lomb L, Malmerberg E, Davidsson J, Nass K, Liang M, Andreasson J, Aquila A, Bajt S, Barthelmess M,

Barty A, Bogan M J, Bostedt C, Bozek J D, Caleman C, Coffee R, Coppola N, Ekeberg T, Epp S W, Erk B, Fleckenstein H, Foucar L, Graafsma H, Gumprecht L, Hajdu J, Hampton C Y, Hartmann R, Hartmann A, Hauser G, Hirsemann H, Holl P, Hunter M S, Kassemeyer S, Kimmel N, Kirian R A, Maia F R N C, Marchesini S, Martin A V, Reich C, Rolles D, Rudek B, Rudenko A, Schlichting I, Schulz J, Seibert M M, Sierra R G, Soltau H, Starodub D, Stellato F, Stern S, Strueder L, Timneanu N, Ullrich J, Wahlgren W Y, Wang X, Weidenspointner G, Wunderer C, Fromme P, Chapman H N, Spence J C H and Neutze R 2012 *Nature Methods* **9** 263

[3] Chapman H N, Fromme P, Barty A, White T A, Kirian R A, Aquila A, Hunter M S, Schulz J, DePonte D P, Weierstall U, Doak R B, Maia F R N C, Martin A V, Schlichting I, Lomb L, Coppola N, Shoeman R L, Epp S W, Hartmann R, Rolles D, Rudenko A, Foucar L, Kimmel N, Weidenspointner G, Holl P, Liang M N, Barthelmess M, Caleman C, Boutet S, Bogan M J, Krzywinski J, Bostedt C, Bajt S, Gumprecht L, Rudek B, Erk B, Schmidt C, Homke A, Reich C, Pietschner D, Struder L, Hauser G, Gorke H, Ullrich J, Herrmann S, Schaller G, Schopper F, Soltau H, Kuhnel K U, Messerschmidt M, Bozek J D, Hau-Riege S P, Frank M, Hampton C Y, Sierra R G, Starodub D, Williams G J, Hajdu J, Timneanu N, Seibert M M, Andreasson J, Rocker A, Jonsson O, Svenda M, Stern S, Nass K, Andritschke R, Schroter C D, Krasniqi F, Bott M, Schmidt K E, Wang X Y, Grotjohann I, Holton J M, Barends T R M, Neutze R, Marchesini S, Fromme R, Schorb S, Rupp D, Adolph M, Gorkhover T, Andersson I, Hirsemann H, Potdevin G, Graafsma H, Nilsson B and Spence J C H 2011 *Nature* **470** 73–U81 ISSN 0028-0836

[4] Vinko S M, Ciricosta O, Cho B I, Engelhorn K, Chung H K, Brown C R D, Burian T, Chalupsky J, Falcone R W, Graves C, Hajkova V, Higginbotham A, Juha L, Krzywinski J, Lee H J, Messerschmidt M, Murphy C D, Ping Y, Scherz A, Schlotter W, Toleikis S, Turner J J, Vysin L, Wang T, Wu B, Zastrau U, Zhu D, Lee R W, Heimann P A, Nagler B and Wark J S 2012 *Nature* **482** 59–62 ISSN 0028-0836

[5] Glover T E, Fritz D M, Cammarata M, Allison T K, Coh S, Feldkamp J M, Lemke H, Zhu D, Feng Y, Coffee R N, Fuchs M, Ghimire S, Chen J, Shwartz S, Reis D A, Harris S E and Hastings J B 2012 *Nature* **488** 603–+ ISSN 0028-0836

[6] David C, Gorelick S, Rutishauser S, Krzywinski J, Vila-Comamala J, Guzenko V A, Bunk O, Farm E, Ritala M, Cammarata M, Fritz D M, Barrett R, Samoylova L, Gruenert J and Sinn H 2011 *Scientific Reports* **1** 57

[7] Nilsson D, Uhlen F, Holmberg A, Hertz H M, Schropp A, Patommel J, Hoppe R, Seiboth F, Meier V, Schroer C G, Galtier E, Nagler B, Lee H J and Vogt U 2012 *Opt. Lett.* **37** 5046

[8] Yumoto H, Mimura H, Koyama T, Matsuyama S, Tono K, Togashi T, Inubushi Y, Sato T, Tanaka T, Kimura T, Yokoyama H, Kim J, Sano Y, Hachisu Y, Yabashi M, Ohashi H, Ohmori H, Ishikawa T and Yamauchi K 2013 *Nature Photonics* **7** 43–47

[9] Schropp A, Hoppe R, Meier V, Patommel J, Seiboth F, Lee H J, Nagler B, Galtier E C, Arnold B, Zastrau U, Hastings J B, Nilsson D, Uhlen F, Vogt U, Hertz H M and Schroer C G 2013 *Scientific Reports* **3** 1633

[10] Yan H and Chu Y S 2013 *J. Synchrotron Rad.* **20** 89–97

[11] Schropp A, Hoppe R, Patommel J, Seiboth F, Uhlén F, Vogt U, Lee H J, Nagler B, Galtier E C, Zastrau U, Arnold B, Heimann P, Hastings J B and Schroer C G 2013 *Proc. SPIE* **8849** 88490R

[12] Lengeler B, Schroer C, Tümmler J, Benner B, Richwin M, Snigirev A, Snigireva I and Drakopoulos M 1999 *J. Synchrotron Rad.* **6** 1153–1167

[13] Lengeler B, Schroer C G, Kuhlmann M, Benner B, Günzler T F, Kurapova O, Somogyi A, Snigirev A and Snigireva I 2004 *Synchrotron Radiation Instrumentation* (*AIP Conference Proceedings* no 705) ed Warwick T, Arthur J, Padmore H A and Stöhr J pp 748–751

[14] Rodenburg J M and Faulkner H M L 2004 *Appl. Phys. Lett.* **85** 4795–4797

[15] Rodenburg J M, Hurst A C, Cullis A G, Dobsen B R, Pfeiffer F, Bunk O, David C, Jefimovs K and Johnson I 2007 *Phys. Rev. Lett.* **98** 034801

[16] Thibault P, Dierolf M, Menzel A, Bunk O, David C and Pfeiffer F 2008 *Science* **321** 379–382

[17] Thibault P, Dierolf M, Bunk O, Menzel A and Pfeiffer F 2009 *Ultramicroscopy* **109** 338–343

[18] Maiden A M and Rodenburg J M 2009 *Ultramicroscopy* **109** 1256–1262

[19] Schropp A, Boye P, Goldschmidt A, Hönig S, Hoppe R, Patommel J, Rakete C, Samberg D, Stephan S, Schöder S, Burghammer M and Schroer C G 2011 *J. Microscopy* **241** 9–12

[20] Giewekemeyer K, Thibault P, Kalbfleisch S, Beerlink A, Kewish C M, Dierolf M, Pfeiffer F and Salditt T 2010 *P. Natl. Acad. Sci. USA* **107** 529–534

[21] Schropp A, Hoppe R, Patommel J, Samberg D, Seiboth F, Stephan S, Wellenreuther G, Falkenberg G and Schroer C G 2012 *Appl. Phys. Lett.* **100** 253112

[22] Hoppe R, Reinhardt J, Hofmann G, Patommel J, Grunwaldt J D, Damsgaard C D, Wellenreuther G, Falkenberg G and Schroer C G 2013 *Appl. Phys. Lett.* **102** 203104

[23] Dierolf M, Menzel A, Thibault P, Schneider P, Kewish C M, Wepf R, Bunk O and Pfeiffer F 2010 *Nature* **467** 436–440

[24] Diaz A, Trtik P, Guizar-Sicairos M, Menzel A, Thibault P and Bunk O 2012 *Phys. Rev. B* **85** 020104(R)

[25] Wilke R N, Priebe M, Bartels M, Giewekemeyer K, Diaz A, Karvinen P and Salditt T 2012 *Opt. Express* **20** 19232–19254

[26] Schropp A, Boye P, Feldkamp J M, Hoppe R, Patommel J, Samberg D, Stephan S, Giewekemeyer K, Wilke R N, Salditt T, Gulden J, Mancuso A P, Vartanyants I A, Weckert E, Schöder S, Burghammer M and Schroer C G 2010 *Appl. Phys. Lett.* **96** 091102

[27] Kewish C M, Guizar-Sicairos M, Liu C, Qian J, Shi B, Benson C, Khounsary A M, Vila-Comamala J, Bunk O, Fienup J R, Macrander A T and Assoufid L 2010 *Opt. Express* **18** 23420–23427

[28] Kewish C M, Thibault P, Dierolf M, Bunk O, Menzel A, Vila-Comamala J, Jefimovs K and Pfeiffer F 2010 *Ultramicroscopy* **110** 325–329

[29] Vila-Comamala J, Diaz A, Guizar-Sicairos M, Mantion A, Kewish C M, Menzel A, Bunk O and David C 2011 *Opt. Express* **19** 21333–21344

[30] Hönig S, Hoppe R, Patommel J, Schropp A, Stephan S, Schöder S, Burghammer M and Schroer C G 2011 *Opt. Express* **19** 16325–16329

[31] Schroer C G, Boye P, Feldkamp J M, Patommel J, Samberg D, Schropp A, Schwab A, Stephan S, Falkenberg G, Wellenreuther G and Reimers N 2010 *Nucl. Instrum. Meth. A* **616** 93–97

[32] Schropp A, Patommel J, Seiboth F, Arnold B, Galtier E C, Lee H J, Nagler B, Hastings J B and Schroer C G 2012 *Proc. SPIE* vol 8504 p 85040F

Journal of Physics: Conference Series **499** (2014) 012005 doi:10.1088/1742-6596/499/1/012005

Setup of an 8 keV laboratory transmission x-ray microscope

S. Baumbach[1], B. Kanngießer[2], W. Malzer[2], H. Stiel[3], S. Bjeoumikhova[4] and T. Wilhein [1]

[1]Institute for X-Optics, RheinAhrCampus Remagen, University of Applied Sciences Koblenz, Joseph-Rovan-Allee 2, 53424 Remagen, Germany
[2]Berlin Laboratory for innovative X-ray Technologies (BliX), Institute for Optics and Atomic Physics, Technical University of Berlin, Hardenbergstrasse 36, 10623 Berlin, Germany
[3]Berlin Laboratory for innovative X-ray Technologies (BliX), Max-Born-Institut, Max-Born-Strasse 2A, 12489 Berlin, Germany
[4]IFG, Institute for Scientific Instruments GmbH, Rudower Chausse 29/31, 12489 Berlin, Germany

E-mail: baumbach@rheinahrcampus.de

Abstract. This article presents a concept and the first results for the setup of an 8 keV laboratory transmission x-ray microscope with a polycapillary optic as condenser at the BliX in Berlin. The incentive of building such a microscope is that the penetration depth for hard x-rays is much higher than in the soft x-ray range, e.g. the water window. Therefore, it is possible to investigate even dense materials such as metal compounds, bones or geological samples. The future aim is to achieve a spatial resolution better than 200 nm.

1. Introduction

Full field laboratory hard x-ray microscopes have been developed in the last decade and the achieved resolution is below 40 nm [1, 2]. One of the challenges of setting up a hard x-ray transmission microscope in the lab is the long focal length of several millimeters of the objective lens, which leads to very long image distances to get a high detector resolution. Additionally, in order to get a high photon flux in the image plane, the numerical apertures of the condenser and the objective lens should match. This leads to further implications for the design of the condenser lens as well as for a suitable detector system with proper spatial resolution to avoid long image distances. In this article we are showing the first results obtained with a polycapillary optic as condender in our laboratory hard x-ray transmission microscope.

2. Experimental Setup

The experiment setup, see figure 1, consists of a copper x-ray tube (A), a polycapillary optic (B), the object within a pinhole (C), a zone plate as objective lens and central stop (D) and a x-ray CCD (E). In the following the different components are described.

2.1. X-ray source

The x-ray source of the laboratory x-ray microscope is a micro focus tube. The tube has a copper target to get the characteristic line emission at 8.05 keV and a source size of about 50 µm

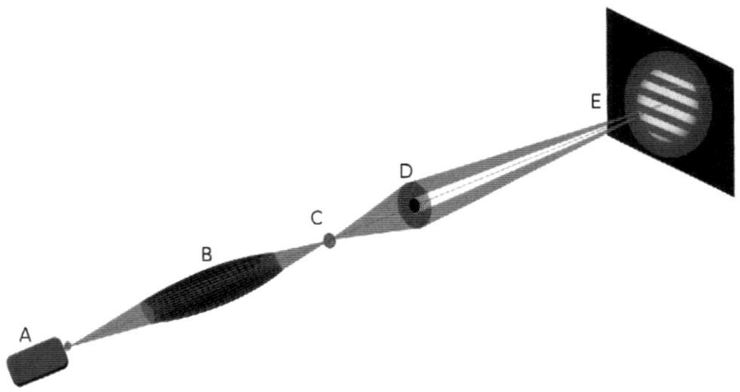

Figure 1. Schematic setup of the transmission microscope.

(FWHM)[3]. Reducing radiation damages to the CCD the maximum acceleration voltage of the tube has been set to 30 kV.

2.2. Condenser

A polycapillary is used as condenser. This optic consists of very thin and bowed hollow glas capillaries acting as wave guide based on total external reflection [4]. All capillaries are pointing to the focal spot resulting in 50 µm spot size (FWHM). A proper condenser should have a high acceptance angle at the entry, a divergence angle at the exit of the capillary matched to the numerical aperture of the zone plate and a high efficiency. In practice, not all conditions can be realized in one device and one has to make a compromise. In our setup the aperture of the condenser optic does not match to the zone plate aperture, i.e. light collected by the condenser does not completely contribute to image formation. But due to the focusing properties of the polycapillary, an increase of the intensity in the object plane is reached and the object is illuminated under different angles. In figure 3 one can see the illumination cone of the capillary optic in a distance of 405 mm to the focal spot.

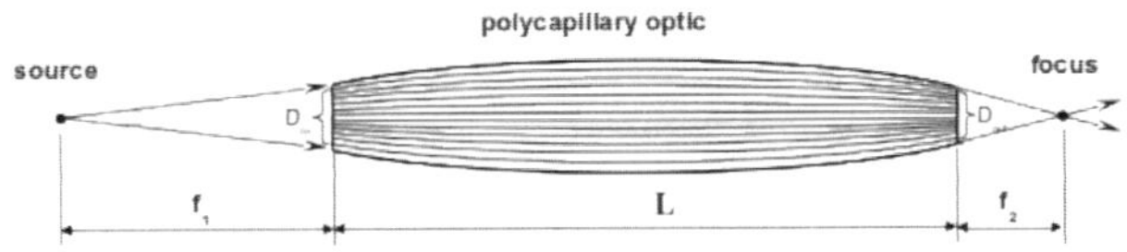

Figure 2. Schematic drawing of a polycapillary optic. Image taken from [4].

Table 1. Parameters of the polycapillary optic. Values are denoted in mm.

f_1	D_{in}	L	f_2	D_{out}
62.3	2.2	45.4	18	1.2

2.3. Objective lens and central stop

A tungsten zone plate with a focal length of about 32.5 mm at 8.05 keV and an outermost zone width of 50 nm is used as objective lens. The zone plate produces a magnified image of the illuminated object on a CCD with 20 µm · 20 µm pixel size. The central stop (CS) is mounted in front of the zone plate and the illumination cone of the polycapillary forms a shadow of the CS on the CCD preventing background illumination. In this shadow the magnified image of the object is formed by the objective. Ideally, the position of the CS is at the exit of the condenser optic. But due to the beam divergency of each capillary a CS on a polycapillary would not form a sharp shadow on the CCD.

Journal of Physics: Conference Series **499** (2014) 012005 doi:10.1088/1742-6596/499/1/012005

3. First results

In the first steps a 30 µm in diameter Pt/Ir-pinhole was used as an object. The resulting image with 108-fold magnification is shown in figure 4. An aluminium aperture with 4 mm in diameter in front of the detector blocks all the light around the image area.

To get a first valuation of the achievable spatial resolution a gold mesh with 5.5 µm bar width and 10.5 µm spaces was used. The resulting 143-fold image is shown in figure 5. For a 143-fold magnification and a pixel size of 20 µm the resulting effecitve pixel size is 140 nm and therefore the Shannon-Nyquist-theorem gives a maximum spatial resolution of $4 \cdot 140\,\text{nm} = 560\,\text{nm}$. The corresponding line plot through the edge of a mesh bar is shown in figure 6 and the measurement from 10% to 90% edge profile gives an estimated spatial resolution of about 580 nm, i.e. the resolution of the system is mainly detector limited.

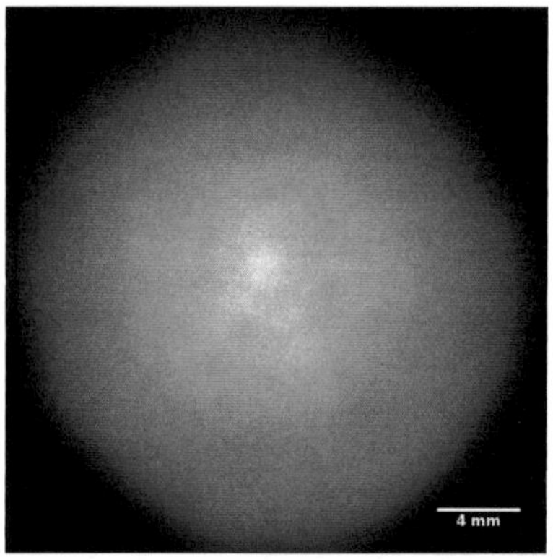

Figure 3. Image of the polycapillary illumination cone. Focal spot to CCD distance is 405 mm.

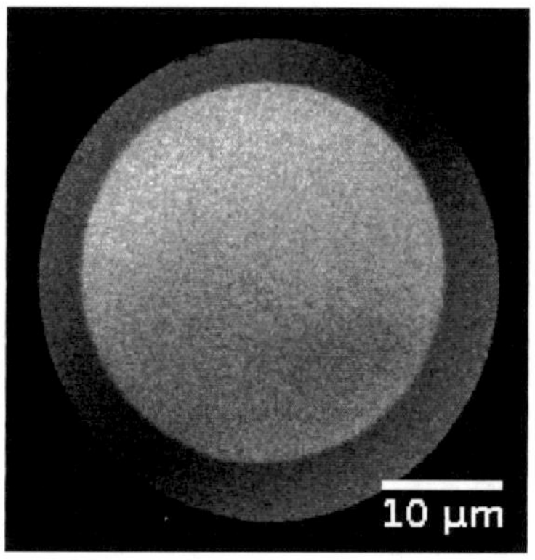

Figure 4. Image of a 30 µm in diameter Pt/Ir-pinhole. 108-fold magnification.

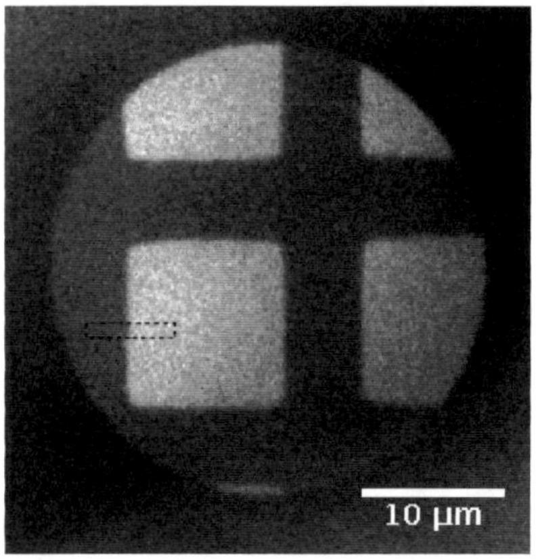

Figure 5. Image of a gold mesh with 1500 line pairs per inch. 143-fold magnification.

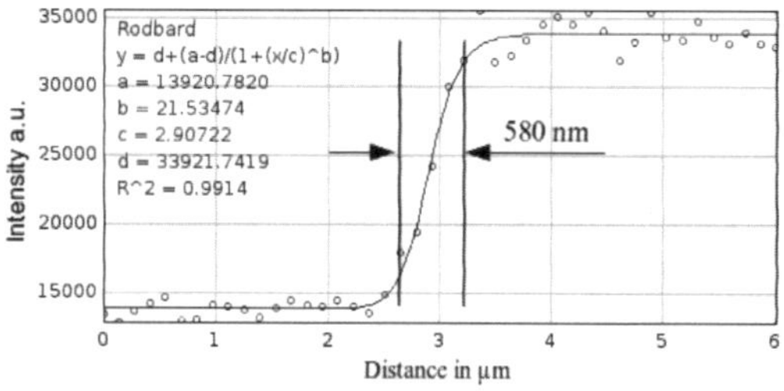

Figure 6. Line plot through the egde of a mesh bar, (black box in figure 5).

4. Summary and Outlook

We have shown that a polycapillary optic can be used as an x-ray condenser in a laboratory 8 keV x-ray transmission microscope. The resolution of the current system is in the sub micro meter region and is still detector limited. A better detector resolution would be achieved with a higher magnification, respectively a longer image distance. But a detector system with smaller pixel sizes, e.g. an image intensifier and a phosphor screen with grain sizes about 2 µm would be the proper choice. In future it is planned to optimize the condenser optic in order to increase the photon flux in the object plane and to get a more adapted output angle to the zone plate. With the described improvements a resolution below 200 nm would be feasible.

Acknowledgements
A special thanks to Mr. Pambos Charalambous (www.ZonePlates.com) for the fabrication of the zone plates.

References
[1] A Tkachuk, M Feser, et al 2006 *Proc. of SPIE Vol. 6318 Developments in X-Ray Tomography V*
[2] Y S Chu,J M Yi,et al 2008 *Applied Physics Letters 92, 103119*
[3] IFG 2012 *X-ray source and power supply manual* IFG Institue for Scientific Instuments GmbH Rudower Chaussee 29/31, 12489 Berlin, Germany
[4] A Bjeoumikhov, S Bjeoumikhova, R Wedell 2006 *Part. Part. Syst. Charact., Vol. 22 p. 384-390*
[5] R Goergl, P Wobrauschek, Ch Streli, H Aiginger, M Benedikt 1995 *X-Ray Spectrometry, Vol. 24, 157-162*

Journal of Physics: Conference Series **499** (2014) 012006 doi:10.1088/1742-6596/499/1/012006

Edge illumination X-ray phase-contrast imaging: nanoradian sensitivity at synchrotrons and translation to conventional sources

P C Diemoz[1,2], M Endrizzi[1], C K Hagen[1], C Rau[3], A Bravin[4], R D Speller[1], I K Robinson[2,5] and A Olivo[1,2]

[1] Dept. of Medical Physics and Bioengineering, UCL, WC1E 6BT London, UK
[2] Research Complex at Harwell, Harwell Oxford Campus, OX11 0DE Didcot, UK
[3] Diamond Light Source, Harwell Oxford Campus, OX11 0FA Didcot, UK
[4] European Synchrotron Radiation Facility, 38043 Grenoble, France
[5] London Centre for Nanotechnology, WC1H 0AH London, UK

E-mail: p.diemoz@ucl.ac.uk

Abstract. Edge illumination is an X-ray phase-contrast imaging technique that can be efficiently applied to both synchrotron radiation and laboratory sources. Its implementation with these two types of setups is here described, and a recently developed method to perform quantitative retrieval of the object attenuation and refraction properties is presented. We report results obtained at two synchrotron radiation facilities and with one of the setups installed in our laboratories at University College London, which show that very high angular sensitivities can be obtained with this technique. The effect of different experimental parameters on the achievable sensitivity is also analyzed. The obtained results will be a useful guide for the design and optimization of future experimental layouts.

1. Introduction

X-ray phase-contrast imaging (XPCi) has been an important branch of X-ray research since the mid nineties [1-4]. The edge illumination (EI) XPCi technique, in particular, has emerged over the last years for its ability to combine high phase sensitivity with a practical and simple implementation [5-11]. EI was first developed and experimentally applied at the Elettra synchrotron radiation (SR) facility, by using a parallel and monochromatic beam [5]. The main idea behind its development was to reproduce the high angular selectivity obtainable with diffraction from a perfect crystal (like in the so-called analyzer-based XPCi technique [1]), however with a simpler setup that does not require a crystal.

Analyzer-based imaging (ABI) is a highly sensitive technique that has been also successfully applied to the imaging of several types of biomedical objects [12,13]. However, it requires strict monochromatization/collimation of the beam, therefore leading to inefficient use of the X-ray flux, and a high mechanical stability, since the crystal orientation has to be controlled with sub-μrad precision. This makes it an excellent tool at SR facilities, but one that is not readily adaptable to conventional x-ray sources.

The development of the EI method arose from the simple observation that a collimated beam is geometrically deflected in the presence of sample refraction, and this deviation can be converted, with a sufficiently large propagation distance, into a detectable spatial shift of the beam. The collimation requires only the use of a one-dimensional (1D) slit, while the displacement of the beam caused by refraction can be sensed by using an absorbing edge set in front of the detector, at a distance on the order of tens of centimetres or a few metres from the sample (figure 1a). The use of such a simple setup, as opposed to the diffraction from a perfect crystal, eliminates the need of a monochromatic beam and of a stringent control of the mechanical stability.

In recent years, the technique has been further developed at University College London (UCL) and its application to laboratory setups based on conventional X-ray tubes was demonstrated [6,7,10,11]. In order to exploit the full size of the two-dimensional (2D) beam, the sample and detector slits are replaced by two masks, made of multiple apertures alternated to absorbing septa (figure 1b). The period of the detector mask is designed so that it matches that of the detector pixels, while the period of the sample mask is calculated so as to account for the magnification caused by the beam divergence. This laboratory implementation of EI has been already successfully applied to the imaging of biological tissues, in particular ex-vivo cartilage and breast samples [14,15].

EI was demonstrated to effectively overcome most of the roadblocks that have so far prevented a widespread deployment of XPCi in standard laboratories, namely the requirements of other XPCi techniques in terms of either beam monochromaticity, small source dimensions, high setup stability, or large doses delivered to the sample. In fact, EI can be applied effectively even with relatively large source sizes, up to at least 100 μm, it is insensitive to broad beam polychromaticity, and robust to mechanical vibrations in the setup. Besides, since the beam is collimated before and not after the sample, unnecessary exposure of the sample is avoided and doses can be kept low (a detailed discussion of dose delivery in lab-based EI goes beyond the scope of this proceedings paper and can be found in [15]).

It should be noted that a more recent XPCi technique, grating interferometry (GI) [16,17], was also implemented with laboratory sources [18]. GI exploits the fractional Talbot effect from a diffraction grating to produce a modulation of the beam intensity: this pattern and the changes to it introduced by the sample are then sensed with the use of a second (absorption) grating placed just before the detector. The application with conventional sources requires the use of a third (absorption) grating near the source, which produces a plurality of self-coherent beamlets [18]. Although there exists a superficial similarity between the EI and GI experimental setups (both use phase or absorption masks/gratings), the physical principles at the basis of the two techniques and their practical implementation are different [19]. For example, one important distinctive feature we would like to mention here is the fact that GI exploits the fractional Talbot effect, which relies on high beam spatial and temporal coherence, while EI is a non-interferometric, intrinsically incoherent method, and can in fact be accurately described with the use of geometrical optics [11,20].

Quantitative retrieval of the object attenuation and refraction maps with EI has also been recently demonstrated, with both synchrotron and laboratory setups, and proven to provide highly accurate results [7,9,11]. Moreover, recent experiments performed at two SR facilities have shown that extremely high angular sensitivity is achievable with this method over a wide range of energies and experimental conditions [9]. The sensitivity of one of the laboratory setups installed at UCL has also been estimated and proven to be at least comparable to results obtained with other XPCi techniques, showing that the simplicity of the setup used by EI does not come at the price of a decreased sensitivity [11].

In the following sections, we describe the application of the EI method and derive a theoretical expression for the technique's sensitivity, dependent upon the used experimental parameters. We then report on some of the main results obtained in our recent experiments carried out at the European Synchrotron Radiation Facility (ESRF, France) and Diamond Light Source (Didcot, UK), and with our laboratory-based setup.

2. The EI technique and its angular sensitivity

As mentioned above, in the EI technique the beam is collimated by a 1D slit before hitting the sample, typically to a few μm or tens of μm, and then analyzed with the use of an absorbing edge (in practice, a slit) in contact with the detector (figure 1a). The adaptation of the method in the case of a spatially extended beam, like that provided by an X-ray tube, is simple, and only requires the use of a pair of sample and detector masks with appropriate periods and aperture sizes (figure 1b). In the normal implementation of EI, the setup is designed so that there is no cross-talk between each sample-detector aperture pair and the adjacent ones. Therefore, in the following we only consider the case of one sample and one detector aperture, as all aperture pairs are independent of the others and the same formalism therefore applies to all in exactly the same way.

The intensity incident on the detector, without the sample, depends on the relative position between each detector aperture and the corresponding beamlet incident on it. In particular, it can be written as the integral of the beam distribution S_{ref} (defined as the number of photons per unit distance), within the two edges of the aperture, i.e. $I_{ref}\left(x_e\right)=\int_{x_e}^{x_e+d}dxS_{ref}\left(x\right)=I_0C\left(x_e\right)$. x_e is the position of the lower edge of the detector aperture, d its dimension, $I_0=\int_{-\infty}^{+\infty}dxS_{ref}\left(x\right)$ is the total intensity of the beam and $C\left(x_e\right)=I_0^{-1}\int_{x_e}^{x_e+d}dxS_{ref}\left(x\right)$ is the so-called illumination curve. The latter represents the fraction of the total beam that enters the aperture: it is a function comprised between 0, when the absorbing part of the mask is totally covering the beam, and a maximum value (close to 1), when the detector and sample apertures are perfectly aligned.

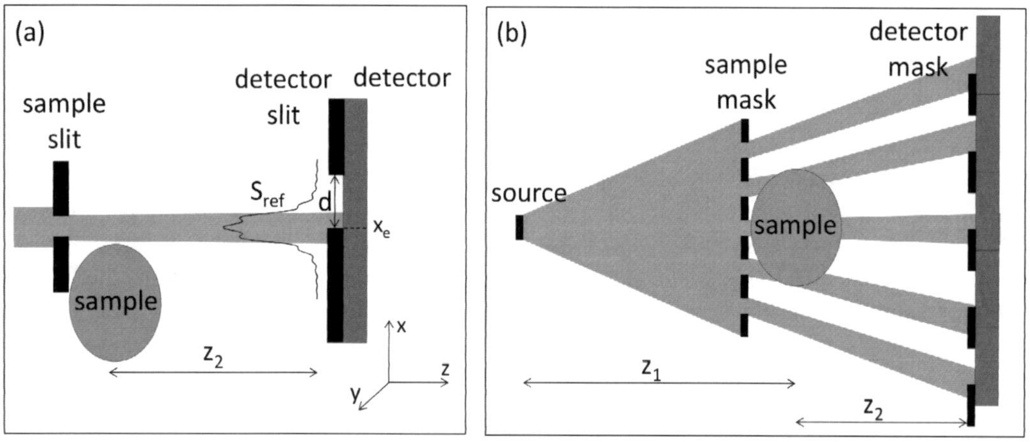

Figure 1. Schematic representation of the edge illumination setup, in the case of (a) a laminar beam and a single pair of sample and detector apertures and (b) a 2-dimensional beam and a pair of sample and detector masks (not to scale). The values of the experimental parameters vary depending on the targeted applications: values for the specific examples provided in this article are given in the text.

When the object is introduced, it can have two effects: beam attenuation, leading to an overall decrease in the beam intensity, and beam refraction (which for samples of interest is typically on the order of several nrad, up to a few μrad). The latter effect causes a spatial shift of the beam at the detector plane equal to $\Delta x=z_2\Delta\theta_x$ (usually in the sub-μm range), where $\Delta\theta_x$ is the component of the refraction angle along the direction x perpendicular to the edge and z_2 is the sample-to detector edge distance.

Mathematically, the intensity on the detector can then be written as [9,11]:

$$I_{obj}\left(x_e\right)=I_0 T_{eff} C\left(x_e-z_2\Delta\theta_{x,eff}\right) \tag{1}$$

where T_{eff} and $\Delta\theta_{x,eff}$ represent the effective beam transmission and refraction angle, calculated as a weighted average over the entire spectrum and the system response function [19]. If two images are acquired at different (arbitrary) positions of the detector aperture, $x_{e,+}$ and $x_{e,-}$, these two components can be separated and quantified (note that in the lab-based implementation of EI it is usually more advantageous to shift the sample mask instead of the detector mask, so that the detector can be kept in a fixed position). In particular, the refraction angle can be calculated by dividing pixel-by-pixel the two images:

$$\Delta\theta_{x,eff}=\frac{1}{z_2}R^{-1}\left(\frac{I_{obj,+}}{I_{obj,-}}\right) \tag{2}$$

where the function R is defined as:

$$R\left(z_2\Delta\theta_{x,eff}\right)\equiv\frac{C\left(x_{e,+}-z_2\Delta\theta_{x,eff}\right)}{C\left(x_{e,-}-z_2\Delta\theta_{x,eff}\right)} \tag{3}$$

and its shape can be computed numerically using the experimental measure of the illumination curve C.

Equation (2) can also be used to estimate, from the noise of the two input images, the uncertainty on the retrieved values of $\Delta\theta_{x,eff}$. Under the assumption of pure statistical noise, in fact, it is possible to demonstrate that [11]:

$$\sigma\left(\Delta\theta_{x,eff}\right)\simeq\frac{\sqrt{C\left(x_{e,+}\right)}}{z_2\sqrt{2TI_0}\left[S_{ref,n}\left(x_{e,+}\right)-S_{ref,n}\left(x_{e,+}+d\right)\right]} \tag{4}$$

where $S_{ref,n}\equiv I_0^{-1}S_{ref}$ is the beam normalized spatial distribution, and where we have assumed that symmetric mask positions are used for the two input images (i.e. the detector mask is shifted with respect to the sample mask of the same amount in each direction). The amplitude of the noise in the refraction image, calculated from equation (4), is a precise indication of the angular sensitivity of an EI setup. In fact, only angles that are larger than the noise level are effectively detectable by the imaging system.

The noise in the processed image, in turn, does not only depend on the noise in the input images, but also on the characteristics of the experimental setup, which define how much the sample refraction is "amplified" to provide the final image signal. In particular, it can be seen from equation (4) that the noise in the refraction image depends upon several setup parameters, including the fraction of the beam falling within the detector aperture (the level of illumination C), the propagation distance and the value of the beam distribution at the two edges of the aperture. This last dependence can be explained in an intuitive way by taking figure 1a as a reference. When the beam is shifted, the number of gained (or lost) photons on the detector, at the lower edge of the aperture, is equal to $\pm\Delta x\cdot S_{ref}\left(x_{e,+}\right)$ (in the small angle approximation). This quantity, however, is partially compensated by the beam lost (or gained) at the edge at the other side of the aperture, i.e. $\mp\Delta x\cdot S_{ref}\left(x_{e,+}+d\right)$. This shows how the dimension of the detector aperture is also an important parameter in the design of an EI setup. In particular, if its size is too small compared to the beam width, then $S_{ref}\left(x_{e,+}+d\right)$ is not negligible, and can negatively affect the amplitude of the detected signal. On the other hand, the period of the masks does not affect the sensitivity, as long as it is large enough to prevent beam spillover between adjacent aperture pairs.

It can be shown [11] that, for a sufficiently large projected source size, the beam distribution can be very well approximated by the geometrical optics (i.e. neglecting the wave properties of the beam and assuming that photons travel along straight lines). In this case, it is simply given by the convolution between the sample aperture projected onto the detector plane and the projected source size. An important consequence of this is that the beam shape has no dependence upon energy. In other words, the optical elements of the setup can be considered to be achromatic.

In order to exemplify the effect of some key geometrical parameters of the setup on the achievable sensitivity, we present in the following the calculation (based on equation (4)) of the relative sensitivity for a variety of different system configurations. Initially, we consider the parameters of one of our lab-based experimental setups: source-to-sample distance z_1 = 1.6 m, z_2 = 0.4 m, sample aperture a = 12 μm, detector aperture d = 20 μm. The misalignment between the masks is set to a 50% level (with, alternatively, the lower or upper edge of the detector aperture being aligned with the centre of the sample aperture for the two input images). While in our current setup the full width at half maximum (FWHM) of the source is about 70 μm, we vary it continuously in the simulation results shown in figure 2, in order to elucidate its effect on the signal amplitude. The calculation is repeated by using smaller (8 μm) and larger (20 μm) sample apertures. In all cases, the sensitivity is constant (at the maximum value) when the projected source size is smaller than the aperture, as in this case the width of each beamlet incident on the detector mask is determined primarily by the aperture size. The sensitivity decreases, instead, when the source starts to dominate over the aperture size. However, this decrease is slow, and even at relatively large source sizes (for example at the value of 70 μm used by our setup) most of the phase sensitivity is still preserved with respect to the ideal case of a point source. When the sample aperture is made narrower, the "plateau" is smaller, but the initial sensitivity is higher as the beam is sharper. The opposite happens when a larger sample aperture is considered.

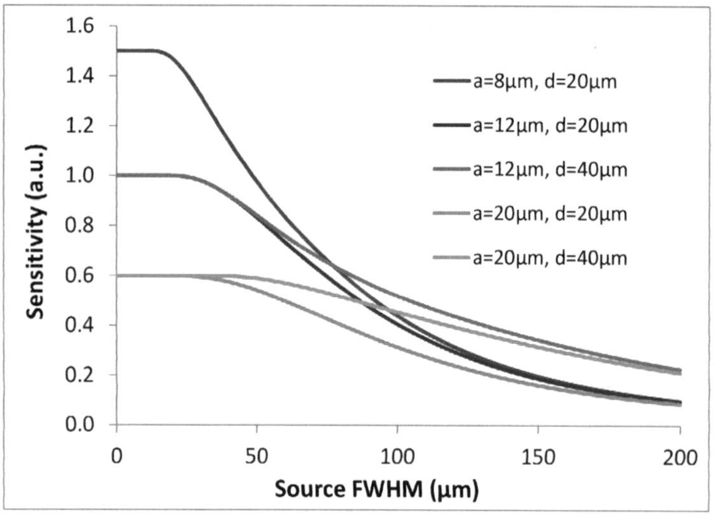

Figure 2. Dependence of the sensitivity upon the dimension of the X-ray source. Curves are calculated using equation (4) for different values of the sample aperture (a) and of the detector aperture (d) (see text for the values of the other experimental parameters).

It is possible to show that part of the signal decrease at large source sizes is actually due to the presence of the second edge of the detector aperture. In fact, in this case the beam incident on it becomes larger than the detector aperture itself. As a consequence, the value of the beam at the upper edge is not negligible, and this leads to a decrease of the signal as discussed above (cf. equation (4)). In order to prove this, we recalculated the sensitivities at a=12 μm and a=20 μm using aperture dimensions increased to d=40 μm. The sensitivity at large source sizes is now drastically improved (see figure 2).

3. Experimental application of EI to synchrotron and laboratory setups

The implementation of the EI technique at SR facilities presents some differences compared to that with a laboratory source. In particular, thanks to the very large source-to-sample distance (sometimes > 100 m), the sample-to-detector distance can be considerably increased (up to several meters), without this making the projected source size (and consequently the beamlet) excessively large [8,9]. As a consequence, the sensitivity can be strongly improved (cf. equation (4)). In addition, since the beam shape is laminar, only one pair of sample and detector slits is usually employed.

The technique was applied using a very high X-ray energy (85 keV) at the ID17 beamline of the ESRF, and using low energy (12 keV) at the I13 beamline (coherent branch) of Diamond. The propagation distances were 6.0 m and 14.7 m, respectively. The aim of these experiments was double: to prove the applicability of EI over a broad energy range, and to experimentally demonstrate that very high sensitivity can be obtained, regardless of the specific characteristics of a given beamline. The experiments proved indeed very high angular resolutions, with a noise in the refraction image of less than 2 nrad with the first setup and of about 15 nrad with the second [9]. This represented a considerable improvement (of up to one order of magnitude better sensitivity) compared to values previously published for other XPCi techniques [9].

The difference between the values obtained in the ESRF and Diamond experiments was due to two main reasons: a higher photon statistics used in the ESRF setup, and a larger broadening of the beam in the Diamond setup, which slightly decreases the sensitivity. The latter effect is due to the beam diffraction, which is not negligible when the projected source size is small, and is more important at lower energies. A consequence of this is that, in SR implementations, the setup can no more be considered achromatic as in the laboratory case.

As an example of the image quality achievable instead in a non-ideal laboratory setup, a refraction image of a wasp, calculated using equation (2), is reported in figure 3. It was obtained using the setup described in the previous section, which is installed in the Radiation Physics laboratories at UCL. The periods of the sample and detector masks are 66.8 μm and 83.5 μm, respectively, while the aperture sizes are 12 μm and 20 μm (more details on the masks design can be found in [11]). The apertures are oriented in the vertical direction, so that the setup is sensitive to beam deflections in the horizontal plane. The two images used as input for the phase retrieval were acquired with a 50% misalignment between the two masks, respectively in the positive and negative directions, at a tube voltage of 35 kVp and with an exposure time of 7 s each. It should be noted that, although the 50% misalignment leads to X-rays intercepted by the detector mask and therefore to a dose increase, this can be avoided by reducing the apertures in the pre-sample mask, as described in [15]. The detector was the ANRAD 'SMAM' amorphous selenium flat panel, with a pixel size of 85 μm. In order to improve the spatial resolution, 8 dithering steps were performed, i.e. the image acquisition was repeated 8 times, each with a sub-period displacement of the sample in the horizontal direction [21] (note that, if a resolution corresponding to the pixel size is already considered to be sufficient, there is no need for any sample dithering, i.e. the sample is not moved at all). The high angular resolution obtained is demonstrated by the detection of faint details in the insect, as for example the structures of the wasp wings and of the chest (figure 3).

The angular sensitivity of this system was estimated as the standard deviation of the values in large regions (500 x 100 pixels^2, outside the sample) of the refraction map. This calculation gave a value of 270±5 nrad, highly homogeneous over the entire field of view [11]. Furthermore, similar calculations performed for other images gave perfectly consistent results.

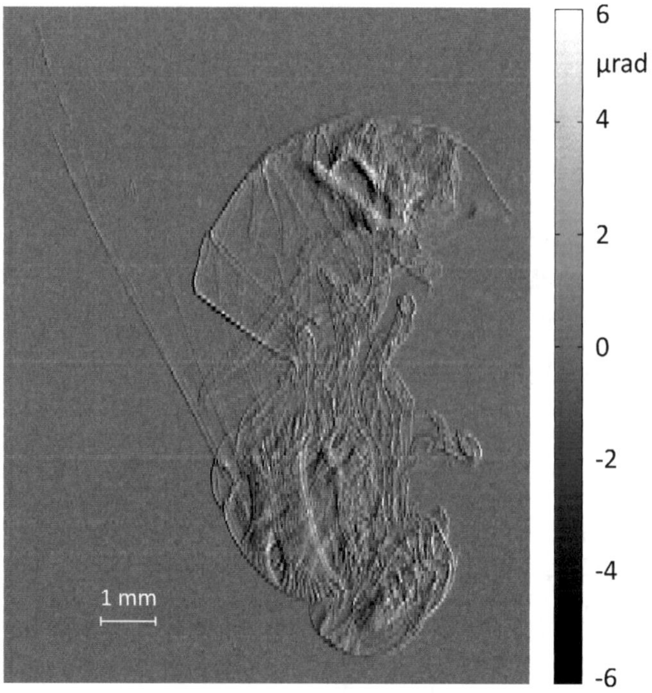

Figure 3. Edge illumination refraction image of a wasp, obtained with our laboratory-based setup (details in text).

The obtained sensitivity is comparable with values previously published for GI. Different implementations of the latter method, in fact, provided sensitivities varying from 100 to 700 nrad [22-24]. However, in these cases, a series of at least 12 images were acquired in order to perform the phase retrieval, instead of only two required by our method when dithering is not needed. Note that the only purpose of sample dithering is to increase the object sampling rate i.e. the spatial resolution, while it has no effect whatsoever on the noise properties and the sensitivity, as the photon statistics in each pixel of the final image is not increased [21]. Conversely, in the case of the 12 "phase stepping" images used in GI, these all contribute to the statistics in the final, processed image, thus reducing the noise approximately by a factor $\sqrt{12}$. Considering that phase sensitivity is defined (here and elsewhere, e.g. [23, 24]) as the standard deviation of the measured angle, it is clear that this is directly influenced by the detected statistics. For example, the best result reported for GI (approximately 110 nrad, [22]) was obtained with 12 phase steps, i.e. by acquiring 12 images with an exposure time of 6.7 s each. By dividing our value of 270 nrad by $\sqrt{6}$ (ratio between the 12 images used in [22] and the 2 used by us, practically with the same exposure time), one obtains roughly 110 nrad, i.e. practically an exact match with the results obtained with GI.

These results demonstrate that, despite the simplicity and practicality of our setup, high angular resolution is obtainable. Besides, we would like to underline the high robustness of the setup with respect to mechanical stability. Environmental vibrations up to a few µm were in fact measured in our laboratory, however these did not affect the quality and the quantitative accuracy of the obtained images. This increased stability is a direct consequence of the large pitch of our masks: it was first observed in the original embodiments of the laboratory prototypes [25], and was recently discussed in more detail [10].

4. Conclusions

The EI method has been successfully applied to both synchrotron and laboratory setups. It was proven to yield unprecedented angular resolution at synchrotrons, thanks to the possibility of using very large propagation distances, on the order of 5-15 m: this enables the amplification of very small refraction angles, which are converted into a detectable displacement of the beam at the detector level. Furthermore, the technique was demonstrated to be efficiently applicable over a broad range of X-ray energies, thus allowing studying a wide variety of different samples.

Thanks to its robustness against increased source dimensions, and to its insensitivity to the beam polychromaticity, EI is also compatible with the use of conventional X-ray tubes. Other key advantages of this setup are the possibility of using very large fields of view, as large masks up to several centimetres can be manufactured, and its resilience to mechanical vibrations. The latter point is particularly important in view of the potential application of the method in normal laboratories and ultimately to clinics. It is also advantageous for the accuracy of the image reconstruction in the cases when more than one image needs to be acquired and processed, like in computed tomography or when phase retrieval is performed.

The implementation of the method in both synchrotron and laboratory setups was discussed, and we have reported on the angular sensitivities obtained in these cases. We have also studied theoretically the dependence of the sensitivity upon various experimental parameters, and in particular the effects of the source dimension, and sample and detector apertures. The developed formalism and the highlighted dependences on the parameters will be key in the design and optimization of future experimental EI setups.

Acknowledgements

This work was supported by the UK Engineering and Physical Sciences Research Council (Grant Nos. EP/G004250/1 and EP/I021884/1). M.E. and P.C.D. are supported by Marie Curie Career Integration Grant Nos. PCIG12-GA-2012-334056 and PCIG12-GA-2012-333990 within the Seventh Framework Programme of the European Union.

References

[1] Ingal V N and Beliaevskaya E A 1995 *J. Phys. D* **28** 2314-7
[2] Snigirev A, Snigireva I, Kohn V, Kuznetsov S and Schelokov I 1995 *Rev. Sci. Instrum.* **66** 5486-92
[3] Wilkins S W, Gureyev T E, Gao D, Pogany A and Stevenson W 1996 *Nature* **384** 35-8
[4] Momose A, Takeda T, Itai Y and Hirano K 1996 *Nat. Med.* **2** 473–5
[5] Olivo A *et al.* 2001 *Med. Phys.* **28** 1610-9
[6] Olivo A and Speller R D 2007 *Appl. Phys. Lett.* **91** 074106
[7] Munro P R T, Ignatyev K, Speller R D and Olivo A 2012 *Proc. Natl. Acad. Sci. USA* **109** 13922-7
[8] Olivo A, Diemoz P C and Bravin A 2012 *Opt. Letters* **37** 915-7
[9] Diemoz P C, Endrizzi M, Zapata C E, Pešić Z D, Rau C, Bravin A, Robinson I K and Olivo A 2013 *Phys. Rev. Lett.* **110** 138105
[10] Millard T P, Endrizzi M, Ignatyev K, Hagen C K, Munro P R T, Speller R D and Olivo A 2013 *Rev. Sci. Instrum.* **84** 083702
[11] Diemoz P C, Hagen C K, Endrizzi M and Olivo A 2013 *Appl. Phys. Lett.* **103** 244104
[12] Coan P, Bamberg F, Diemoz P C, Bravin A, Timpert K, Mutzel E, Raya J G, Adam-Neumair S, Reiser M F and Glaser C 2010 *Invest. Radiol.* **45** 437-44
[13] Zhao Y *et al.* 2012 *Proc. Natl. Acad. Sci. USA* **109** 18290-4
[14] Marenzana M, Hagen C K, Das Neves Borges P, Endrizzi M, Szafraniec M B, Ignatyev K and Olivo A 2012 *Phys. Med. Biol.* **57** 8173-84
[15] Olivo A *et al.* 2013 *Med. Phys.* **40** 090701

[16] David C, Nöhammer B, Solak H and Ziegler E 2002 *Appl. Phys. Lett.* **81** 3287-9
[17] Momose A, Kawamoto S, Koyama I, Hamaishi Y, Takai K and Suzuki Y 2003 *Jpn. J. Appl. Phys.* **42** L866-8
[18] Pfeiffer F, Weitkamp T, Bunk O and David C 2006 *Nat. Phys.* **2** 258-61
[19] Munro P R T and Olivo A 2013 *Phys. Rev. A* **87** 053838
[20] Munro P R T, Ignatyev K, Speller R D and Olivo A 2010 *Opt. Express* **18** 4103-17
[21] Ignatyev K, Munro P R T, Speller R D and Olivo A 2011 *Rev. Sci. Instrum.* **82** 073702
[22] Revol V, Kottler C, Kaufmann R, Straumann U and Urban C 2010 *Rev. Sci. Instrum.* **81** 073709
[23] Thüring T, Modregger P, Hämmerle S, Weiss S, Nüesch J and Stampanoni M 2012 *AIP Conf. Proc.* **1466**, 293-8
[24] Thüring T, Hämmerle S, Weiss S, Nüesch J, Meiser J, Mohr J, David C and Stampanoni M 2013 *Proc. of SPIE* **8668** 866813
[25] Olivo A, Ignatyev K, Munro P R T and Speller R D 2011 *Appl. Opt.* **50** 1765-9

X-ray Full Field Microscopy at 30 keV

F. Marschall*[1], A. Last[1], M. Simon[2], M. Kluge[1], V. Nazmov[1], H.Vogt[1], M. Ogurreck[3], I. Greving[3] and J. Mohr[1]

[1]Institute of Microstructure Technology, Karlsruhe Institute of Technology (KIT/IMT)
[2]PI miCos GmbH, Freiburger Straße 30, 79427 Eschbach
[3]Helmholtz-Zentrum Geesthacht, Institute of Materials Research

felix.marschall@kit.edu

Abstract. In our X-ray full field microscopy experiments, we demonstrated a resolution better than 260 nm over the entire field of view of 80 µm x 80 µm at 30 keV. Our experimental setup at PETRA III, P05, had a length of about 5 m consisting of an illumination optics, an imaging lens and a detector. For imaging, we used a compound refractive lens (CLR) consisting of mr-L negative photo resist, which was fabricated by deep X-ray lithography. As illumination optics, we choose a refractive rolled X-ray prism lens, which was adapted to the numerical aperture of the imaging lens.

1. Introduction

Hard X-ray full field microscopy allows high resolution transmission imaging of samples, which are opaque for visible light. This is often required in materials science and microstructure technology. Today most of the X-ray microscopes installed at synchrotron radiation sources or with X-ray tubes are based on Fresnel zone plates. A reasonable resolution can only be achieved for energies up to about 20 keV, due to the limitation in fabrication of high aspect ratio zone plates with an outer ring widths in the submicrometer range. Microscopes, using refractive lenses for hard X-rays, have been built up at synchrotron beam lines [1] and tube sources [2, 3]. Due to several reasons, full field microscopes for hard X-rays were strongly limited in resolution or required large distances between the optical components. Here we realized an X-ray full field microscope for 30 keV, with higher resolution and shorter setup length, using compound refractive lenses (CLR) out of mr-L negative photo resist. In our very first experiment we achieved a resolution better than 260 nm over a field of view of 80 µm x 80 µm with a setup of around 5 m length. Our results show that this experimental setup has a high potential to revolutionize the field of hard X-ray full field microscopy, opening up new possibilities for studying materials in research and industry.

2. Microscope setup

The microscope was setup at beamline P05 at PETRA III. A sketch of the setup is shown in Figure 1. The microscope consists of an illumination optics, an adjustable diaphragm, an objective lens and a detector. A biconcave parabolic compound refractive lens was used as imaging optics and a rolled X-ray prism lens for the illumination optics. As detector we choose a high resolution system using a LSO-scintillator and a magnifying optics for imaging the fluorescent light on the CMOS-detector.

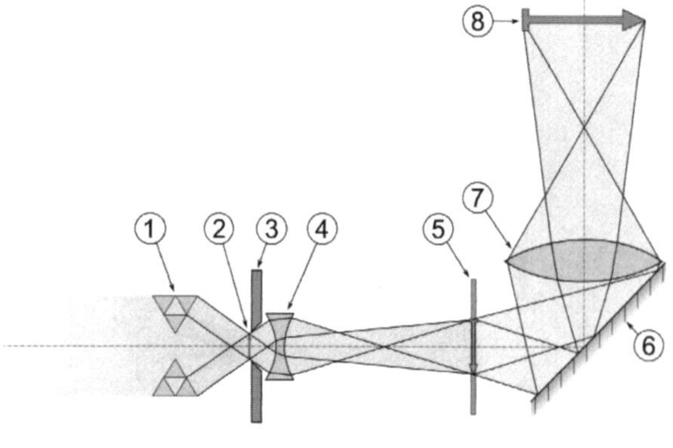

Figure 1: Setup of an X-ray full field microscope with refractive lenses for synchrotron beam lines, consisting of 1) condenser lens, 2) sample, 3) diaphragm, 4) objective lens, 5) scintillator, 6) 45° mirror, 7) microscope objective for visible light, 8) CMOS-detector; the path of the X-rays until the scintillator is drawn in green, the path of the visible light starting at the scintillator is drawn in yellow

In order to achieve the optimum performance of the microscope the magnification factor needs to be adapted to the achievable resolutions in the different imaging planes. For a theoretical resolution of the objective lens of 60 nm and a resolution of the scintillator of 1.5 µm the magnification in the X-ray part of the microscope has to be at least 26. This ensures that all details resolved by the objective lens are also visible on the scintillator. Additionally, selecting only the necessary X-ray magnification for highest resolution, allows the length of the microscope to be as small as possible.

In the same way, the size of the pixels of the CMOS-detector defines the magnification of the visible light objective in front of the detector. For resolving one unit of line and space at least 2.5 pixels, better 4 pixels or more, are needed. Therefore, a magnification of 20 is needed to image 1.5 µm line and space on 4 pixels with a size of 7.3 µm.

For an objective lens with a focal length of 90 mm the sample distance was set to 95 mm and the image distance was 2500 mm which results in the selected magnification of 26. In combination with an illumination lens, which had an illumination distance of 2.5 m, the overall length of the microscope was about 5 m.

3. Objective lens

A compound refractive lens was used as imaging lens. The lens had a focal length of 90 mm and an entrance aperture of 71 µm. It was fabricated by deep X-ray lithography out of mr-L negative photo resist [4] at KIT/IMT. The lithography was done at the X-ray lithography beamline Litho 2 at ANKA. Deep X-ray lithography allows fabricating submicron structures with very high precision [5].

However, lithography allows only the fabrication of linear structures; rotational symmetric lenses cannot be produced by this technique. To solve this problem, the refraction for horizontal and vertical direction is separated and effected by different surfaces, which are fabricated by two separate lithographic exposures. An alternating crossed arrangement of horizontal and vertical lens elements however allows 2D-imaging. In order to avoid astigmatism, the number of lens elements for focusing in one direction is increased by one, with the first and the last element having only half of the refractive power than the other elements. Thus, the principal planes for both directions match each other and the focal length for both directions is equal.

The lens we used in our experiment consists of 249 biconcave parabolic lens elements in total, which means 124 biconcave parabolic lens elements in horizontal and 125 vertical direction. The radius of curvature r in the apex of each parabola is 6 µm. The distance d between the two surfaces of one lens element is 7 µm, the distance a between two lens elements is 30 µm (see Figure 2). The total length of the lens, which is a result of the number of lens elements N, the length L of the lens elements and the distance a between two lens elements, is limited to 60 mm due to the current fabrication process at KIT/IMT. Therefore, the aperture of the lens is geometrically limited to 70 µm, if the aperture of all lens elements is equal.

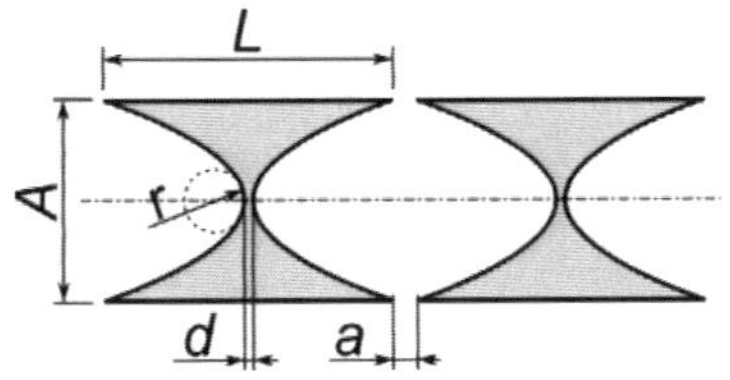

Figure 2: Layout of a parabolic compound refractive X-ray imaging lens (r – radius of curvature in the apex, A – aperture, d – smallest distance between to surfaces of one lens element, a – distance between two lens elements, L – length of one lens element)

The absorption limited aperture of the lens, also called effective aperture [6], is 78 µm. This value is larger than the geometrically possible aperture, which is limited because the overall length of the X-ray lens is limited. Optimizing the apertures of the lens elements to the application is beneficial for achieving best results in each application. An objective lens for full field microscopy has to be optimized in order to achieve a homogeneous and high image quality. Thus, it is important that the lens has a high numerical aperture for all points of the entire field of view.

The result of the optimization is a lens with an aperture varying from 71 µm in the entrance plane, to 66 µm in the middle and to 82 µm in the exit plane [7]. Due to this shape of the lens the resolution increases over the entire field of view. Especially at the borders of the field of view the gain is significant, resulting in a more homogeneous resolution and brightness of the image. For the optimized lens a theoretical resolution of 60 nm in the center and 78 nm at the borders of a field of view of 100 µm x 100 µm was calculated. The average transmission of the lens is 61 %.

To fabricate a lens with horizontal and vertical elements, which are tilted by 90° with respect to each other, two X-ray lithographic exposures with +45° and -45° are needed. The absorber mask used for the exposure consists of a titanium membrane with 2.7 µm thickness, carrying gold absorbers with 20 µm thickness. As the total size of the membrane is 23 mm x 63 mm, the membrane is fragile and avoiding bending of the membrane is very challenging. Due to the 45° exposures bending of the membrane of the absorber mask is directly imaged to the substrate in the lithography step, which results in an incorrect alignment of the lens elements on the substrate. For the lens we used in our experiment we found that the maximum deviation from the ideal position was ± 7 µm. Therefore, the size of the usable aperture was reduced by 14 µm and consequently a theoretical resolution of only 80 nm in the center of the field of view would have been achievable.

4. Condenser lens

Synchrotron beamlines offer a nearly parallel beam. Using imaging lenses for sample illumination results in a small focal spot size, which is sufficient for illuminating a very small area on the sample but not for illuminating an expanded field of view homogeneously (see Figure 3a). Illuminating a larger area by defocusing results in a significantly decreased number of photons on each point of the field of view. Additionally, defocusing reduces the angular variance in illumination directions. While the overall angular spread is constant, each individual point in the field of view suffers from a reduced angular spread (see Figure 3b). So, for illumination an expanded field of view, imaging lenses are not the best choice at synchrotron beamlines. Thus we choose a rolled X-ray prism lens (RXPL) [8] as illumination optics, which provides an homogeneous illumination over an extended field of view (see Figure 3c).

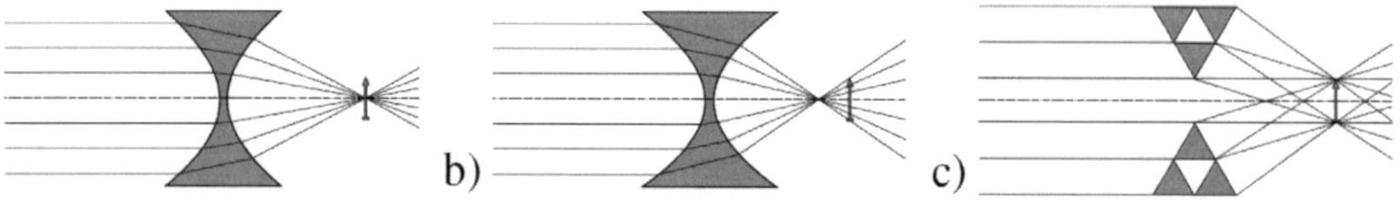

Figure 3: Schematics of illumination optics with parallel incoming rays: a) using an imaing lens focused on the sample, b) using an imaging lens not focused on the sample, c) using an prismlens

Rolled X-ray prism lenses consist of rolled up structured foil, which results in a huge number of layers each containing multiple prisms (see Figure 4). The prisms just change the direction of an incoming parallel beam, not its divergence. Besides the refractive index, the direction of the outgoing ray is determined by the number of prisms in one layer, which can be chosen freely by choosing the shape of the foil before rolling it up. Thus, the incoming beam is cut into sections by each row illuminating a defined part of the sample under a defined direction. As a result each point of the sample is illuminated from multiple directions, which is advantageous in full field microscopy. Additionally, the photon flux on the sample is increased.

The condenser we choose for our experiments had a working distance of 2.5 m, an aperture of 2 mm and a theoretical transmission of 88 %. The lens was specifically designed for illuminating the entire field of view with a divergence tailored to the objective lenses acceptance angle. Thus, we achieved a high brightness of the image and consequently a short exposure time of a few seconds.

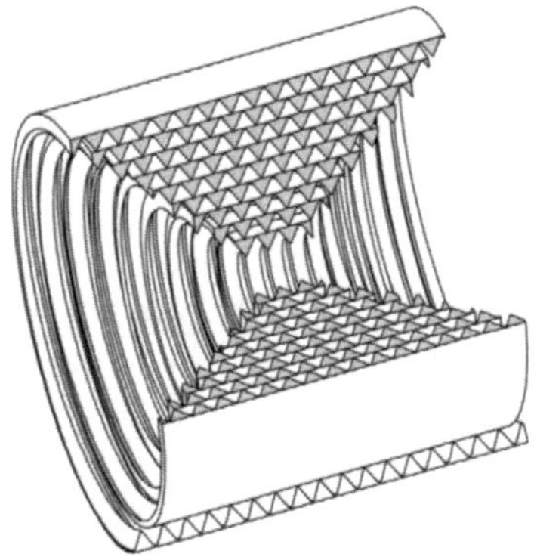

Figure 4: Schematic of a rolled X-ray prism lens

However, rolling up a foil with 10 μm thickness is a very complicated process, and a total avoidance of fabrications defects is highly challenging. We found, that the illumination is not yet completely homogeneous. However our experiment proves that this can be easily compensated for, by dividing the image by its flat field.

5. Experimental results

Our very first experiments for full field microscopy at 30 keV took place at PETRA III, P05 [9]. The 6.8 m long granite optical bench with several high precision parallel kinematic stages mounted on sliders is an excellent basis for full field microscopy. The distance between condenser lens and scintillator was around 5 m for an X-ray optical magnification of 26. As detector we used an 11 μm thick LSO-scintillator combined with a 45°-mirror, a 10-fold or 20-fold microscope objective, and a Nikon D4 camera.

Figure 5: X-ray micrograph of an Xradia nano test pattern at 30 keV divided by its flat field (30 µm outer diameter of Siemens star)

With our setup we have been able to demonstrate a resolution better than 260 nm over the entire field of view of 80 µm × 80 µm, which means that the smallest visible structure details with a width of 130 nm had a periode of 260 nm. Figure 5 shows a detail of a micrograph of an Xradia nano test pattern divided by its flat field. The exposure time for this image was 30 s. The gold layer of the sample was only 500 nm thick, so the transmission was 97.5 % at 30 keV, which is challenging but sufficient for absorption contrast imaging.

6. Conclusion

Our concept based on polymer compound refractive lenses as objective lenses and rolled prism lenses as illumination optics allows building up fast and compact high energy microscopes with high resolution at synchrotron beamlines for ambient conditions. With our microscope, using compound refractive lenses out of mr-L resist, structure details down to 130 nm line and 130 nm space are visible. So, the resulting resolution is 260 nm with a field of view of 80 µm × 80 µm. Our theoretical calculations show that resolutions of below 60 nm are achievable with an optimal set of lenses. As several improvements are currently implemented in the lens fabrication process, we expect approaching this limit in future experiments. The results demonstrate a high potential to make hard X-ray full field microscopy available for materials analysis.

Acknowledgements

The authors thank the Karlsruhe Nano Micro Facility (KNMF), a Helmholtz Research Infrastructure for the possibility to fabricate the lens with X-ray lithography. The work was performed within the Helmholtz Virtual Institute of New X-ray Analytic Methods in Material Science (VI-NXMM).

References

[1] B. Lengeler et al, *A microscope for hard X-rays based on parabolic compound refractive lenses*, Applied Physics Letters, 1999

[2] C. K. Gary et al, *High resolution X-ray microscope*, Applied Physics Letters, 2007

[3] Y. Kohmura et al, High spatial resolution hard X-ray microscope using X-ray refractive lens and phase contrast imaging experiments, Nuclear Instruments and Methods in Physics Research, 2001

[4] E. Reznikova et al, J. Phys.: Conf. Ser. 186, 012070, 2009

[5] V. Saile, *LIGA and its Applications*, Wiley-VCH Verlag GmbH & Co. KGaA, Weinheim, 2008

[6] V. G. Kohn, *An exact theory of imaging with a parabolic continuously refractive X-ray lens*,

Journal of Experimental and Theoretical Physics, 2003
[7] F. Marschall et al, *Taille-lenses: design of aperture-optimised refractive lenses for hard X-ray full field microscopy*, in preparation
[8] H. Vogt et al, Proc. SPIE 8076, 80760I, 2011, doi:10.1117/12.890037
[9] M. Ogurreck et al, *The nanotomography endstation at the PETRA III imaging beamline*, Journal of physics / Conference Series, 2013

Spatial characterization of the focus produced by an EUV Schwarzschild objective

Johannes Ewald[1], Marek Wieland[2], Thomas Nisius[1], Lars Henning[1], Torsten Feigl[3], Markus Drescher[2], and Thomas Wilhein[1]

[1]Institute for X-Optics, University of Applied Sciences Koblenz, Joseph Rovan Allee 2, 53424 Remagen, Germany

[2]Institute for Experimental Physics, University of Hamburg, Luruper Chaussee 149, 22761 Hamburg, Germany

[3]optiX fab., optiX fab GmbH, Hans-Knoell-Straße 6, 07745 Jena, Germany

E-mail: ewald@rheinahrcampus.de

Abstract. Schwarzschild objectives are used in the EUV spectral range because of their large aperture, high mechanical stability and excellent achromaticity. The large aperture results in a small, theoretically diffraction limited focus diameter with ideal values of below 200 nm with the current configuration. We employed a zone plate with matched numerical aperture (0.19) to image the focus onto an X-ray CCD camera. Emission from high harmonic generation and a liquid-jet laser-plasma were used as light sources. Images at magnifications of about 150-fold were acquired and focus diameters of 300 nm (FWHM) were observed.

1. Introduction

To examine non-linear interaction of electromagnetic radiation with matter at short wavelengths, focused beams from free-electron lasers (FELs) can reach the required peak irradiance levels where effects like multiphoton ionization takes place [1]. Methods to analyze spot-sizes of higher-harmonic generation (HHG) sources with a spatial resolution of 1 - 2 µm via knife-edge and fluorescence screen have been presented before [2, 3]. To characterize the ability of the objective to focus EUV radiation using the specified wavelength of 13.5 nm, raytracings and focusing experiments were conducted. After using a He-Ne laser to establish an alignment strategy, experiments using the emission from high harmonic generation were conducted due to the similar beam geometry compared to a free electron laser. In order to examine the focus

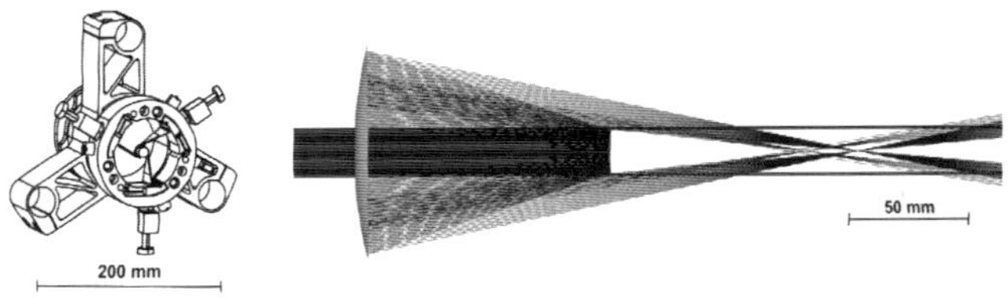

Figure 1. Dimensions and raytracing of a parallel beam through the objective.

profile with a high spatial resolution, an imaging method using a zoneplate as second optic was chosen. By imaging the focus of the Schwarzschild objective with a zoneplate optic on a CCD, one can avoid using, potentially more complex, indirect methods like scanning suitable resolution targets. The practical use of spatial characterization of EUV laser-plasma sources with zoneplates has been shown before [4]. In the case of a strongly demagnified source using the Schwarzschild objective, the requirements on the spatial resolution of the setup increase.

2. Objective properties and setup

The experimental setup is shown in Figure 2. It consists of the objective mounted on a motorized stage. The zoneplate is mounted on an encoded two-axis piezo scanner to achieve the necessary precision for the lateral positioning. Both optics can be moved in the z-direction to focus the imaged spot on the CCD camera. The same experimental setup was used at a liquid jet laser-plasma source at the Institute for X-Optics in Remagen, Germany [5].

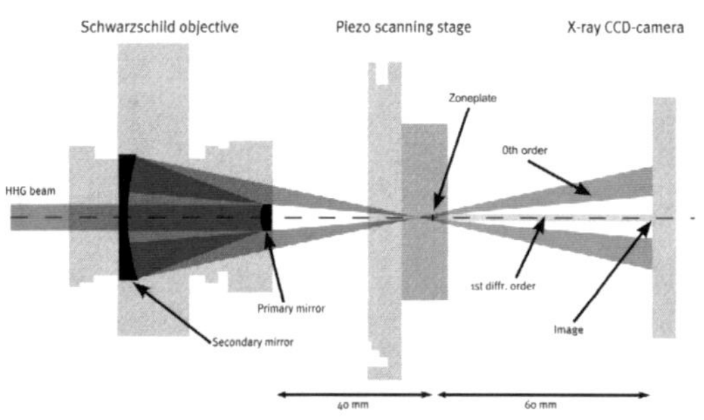

Figure 2. Experimental setup and geometry used to image the focal spot

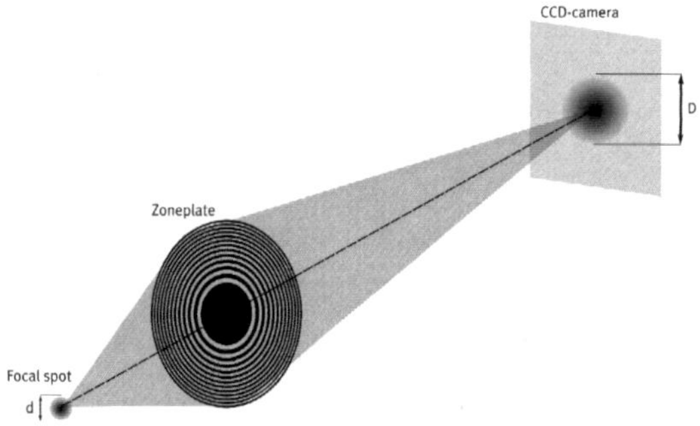

Figure 3. Principal setup to image an X-ray focus on a CCD with a zoneplate

The different components of the setup were mounted in multi-purpose vacuum chambers with piezo driven three-axis positioners. Due to the weight of the Schwarzschild objective, a separate mount allowing six degrees of freedom was developed and manufactured. The focused beam results in a spot and its magnified image is recorded with an X-ray CCD camera. This arrangement is shown in Figure 3. The Schwarzschild objective has an effective focal length of 27 mm, which results in a distance of 42 mm between the last surface of the objective and the focal plane. The focal length of the zoneplate at the wavelength of 13.5 nm is 400 µm. As shown in Figure 2 the short distances involved allow for a very compact setup.

3. Experimental Results

Using the results from the raytraces shown in Figure 1, an alignment strategy using a He-Ne laser was developed. By aligning the concentric images from the different reflection orders, the mirrors of the objective could be correctly positioned relative to each other. After that, only minimal adjustments with the motorized mirror holders were necessary during the actual in-vacuum experiment. The procedure to focus the imaged spot on the X-ray CCD camera is shown in

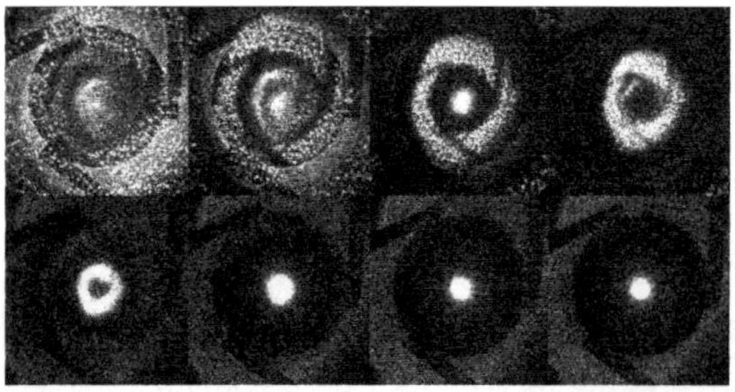

Figure 4. Montage of images taken during the process of finding the optimal position of the zoneplate to the objective in order to focus the image on the CCD

Figure 4. By changing the distance of the zoneplate relative to the spot and camera, the hollow cone of the beam close to the focal spot could be observed. Due to the high numerical aperture of the optics (NA = 0.19), even small lateral movements of the zoneplate could remove the image from the detector field of view. As a result only a small magnification of 150 was chosen, which led to an effective pixel size of 86 nm. Images were taken at a high harmonic generation and a

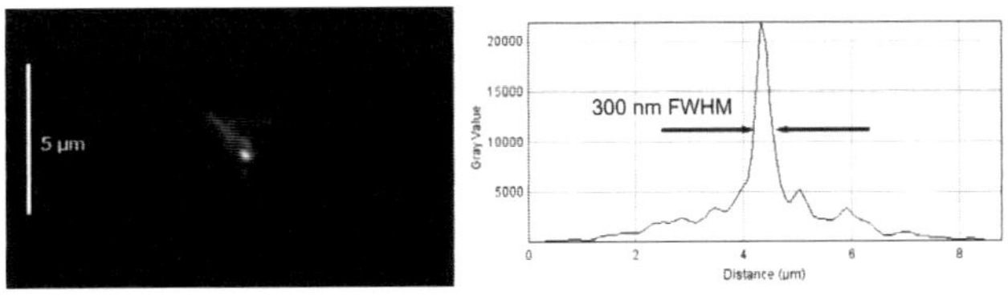

Figure 5. Image of the focused HHG beam with a spot diameter (FWHM) of below 0.5 µm (Magnification = 150, Exposure time = 30 s, Laser repetition rate = 1 kHz)

laser induced laser-plasma source. Due to the optimized layout of the objective for the use at FLASH, small focus sizes could only be achieved with the high harmonic generation source with its similar, parallel beam geometry. One resulting image of the focused beam is shown in Figure 5.

4. Discussion

In experiments at laboratory soft X-ray sources, minimal imaged spot diameters of 300 nm were observed. The imaged spot is surrounded by a plateau of about 4 µm diameter as can be seen in Figure 5. If this is an effect of an distorted incoming wavefront or due to the imaging system is not clear. Due to the scheduled experiments at FLASH (DESY), procedures were evaluated and tested to prealign the objective relative to the vacuum chamber with a pilot laser. This should minimize the amount of beamtime needed to align the system. As a next step, additional to the direct imaging of the spot size with a zoneplate, an ion microscope will be combined with the current setup. The charge states of Xe ions in the focus of the objective are related to the intensity in this area and can be resolved spatially with the ion microscope setup [6]. Both results combined will help to calculate the actual number of photons in the focus area contributing to experiments using the Schwarzschild objective.

References

[1] R Guichard, M Richter, JM Rost, U Saalmann, AA Sorokin, and K Tiedtke. Multiple ionization of neon by soft x-rays at ultrahigh intensity. *arXiv preprint arXiv:1302.0021*, 2013.

[2] L Le Déroff, P Salieres, and B Carré. Beam-quality measurement of a focused high-order harmonic beam. *Optics letters*, 23(19):1544–1546, 1998.

[3] H Mashiko, A Suda, and K Midorikawa. Focusing multiple high-order harmonics in the extreme-ultraviolet and soft-x-ray regions by a platinum-coated ellipsoidal mirror. *Applied optics*, 45(3):573–577, 2006.

[4] U Vogt, M Lindblom, P Charalambous, B Kaulich, and T Wilhein. Condenser for Koehler-like illumination in transmission x-ray microscopes at undulator sources. *Optics letters*, 31(10):1465–1467, 2006.

[5] L Henning. Charakterisierung der Spotgröße eines Schwarzschild Objektivs im EUV Bereich. diploma thesis, HS Koblenz, 2013.

[6] Marek Wieland, Thomas Gebert, Dimitrios Rompotis, Ulrike Frühling, Johannes Ewald, Thomas Nisius, Thomas Wilhein, Elke Plönjes, and Markus Drescher. Following non-linear x-ray physics in the time-domain proposal f-20110346/progress report 2012.

Journal of Physics: Conference Series **499** (2014) 012009 doi:10.1088/1742-6596/499/1/012009

Time-resolved soft X-ray microscopy of magnetic nanostructures at the P04 beamline at PETRA III

P Wessels[1], J Ewald[2], M Wieland[1], T Nisius[2], G Abbati[2], S Baumbach[2], J Overbuschmann[3], A Vogel[4], A Neumann[4], J Viefhaus[5], H P Oepen[4], G Meier[4], T Wilhein[2] and M Drescher[1]

[1] Institut für Experimentalphysik, University of Hamburg, Luruper Chaussee 149, 22761 Hamburg, Germany

[2] Institute for X-Optics, University of Applied Sciences Koblenz, RheinAhrCampus, Joseph-Rovan-Allee 2, 53424 Remagen, Germany

[3] Center of Advanced European Studies and Research (caesar), Project Group Electron Microscopy and Analytics (EMA), Ludwig-Erhard-Allee 2, 53175 Bonn, Germany

[4] Institut für Angewandte Physik, University of Hamburg, Jungiusstraße 11, 20355 Hamburg, Germany

[5] Deutsches Elektronen-Synchrotron DESY, Notkestraße 85, 22607 Hamburg, Germany

E-mail: philipp.wessels@desy.de

Abstract. We present first time-resolved measurements of a new mobile full-field transmission microscope [1] obtained at the soft X-ray beamline P04 at the high brilliance synchrotron radiation source PETRA III.

A nanostructured magnetic permalloy ($Ni_{80}Fe_{20}$) sample can be excited by the magnetic field of a 400 ps full width at half maximum (FWHM) long electric current pulse in a coplanar waveguide. The full-field soft X-ray microscope successively probes the time evolution of the sample magnetization via X-ray magnetic circular dichroism (XMCD) [2] spectromicroscopy in a pump-probe scheme by varying the delay between pump and probe pulses electronically. Static and transient magnetic fields of a permanent magnet and a coil are available in the sample plane to reset the system and to provide external offset fields.

The microscope generates a flat-top illumination field of 20 µm diameter by using a grating condenser [3] and the sample plane is directly imaged by a micro zone plate with 60 nm resolution onto a 2D gateable X-ray detector to select the particular bunch in the storage ring that contains the dynamic information. The setup is built into a mobile endstation vacuum system with in-house developed three-axis piezo motorized stages for high accuracy positioning of all microscopy-components inside the chambers.

1. Motivation and introduction

Besides the fascination of exploring magnetism at its fundamental time- and length-scales itself, research in this direction motivates many applications in modern data storage technology. The demand for smaller and faster storage devices has to be based on primal knowledge of magnetization processes happening at time scales from picoseconds to femtoseconds and length scales from nanometers [4–6] to single atoms [7] at once. The challenging investigation of magnetic systems influenced by femtosecond laser pulses [8–11] or short and confined magnetic field pulses [12–14] is of topical research interest.

In order to contribute to the research done in this direction we have set up a new mobile and flexible time-resolved magnetic transmission X-ray microscope (TR-MTXM) at the soft X-ray beamline P04 at PETRA III. The setup is able to investigate the magnetization dynamics of previously excited nanoscale objects by X-ray magnetic circular dichroism (XMCD) spectromicroscopy [2, 15, 16].

2. Experimental Setup

The schematic setup of the full-field transmission X-ray microscope (TXM) is outlined in figure 1 and also described in [1]. The circular polarized soft X-ray pulses of the P04 beamline at photon energies of the transition metal L edges are focused down to a flat-top spot of 20 µm diameter for Köhler-like illumination by a grating condenser [3] with included gold central stop and a focal length of $f_c = 34$ mm at 852 eV photon energy. A micro zone plate of 80 µm diameter and outermost zone width of 50 nm images the sample plane located at a distance $g = 2.75$ mm onto a 2D gateable pixel detector at $b = 1930$ mm behind the zone plate corresponding to a magnification of $M \approx 700$.

The time resolution is implemented in a pump-probe approach where the sample is excited by a short magnetic field pulse emerging from current pulses in the conductor of a waveguide. The current pulses of 400 ps full width at half maximum (FWHM) duration are created by a picosecond pulse generator synchronized to an external clock of the storage ring. The delay Δt between pump and probe pulses is electronically tunable with a jitter of $\sigma < 30$ ps. In "timing mode" the storage ring PETRA III is filled with 40 electron bunches and delivers a light pulse every 192 ns for probing the sample. However, the sample can only be excited every 7.7 µs matching the revolution frequency of the synchrotron due to the high intensity of the pump pulses. In consequence, this means that only 1 of 40 bunches contains a transient signal and the unpumped bunches have to be filtered. For this purpose an in-house developed fast gateable X-ray detector is used: The microscope image is projected onto a P46 phosphor screen and imaged via a relay optic consisting of two fast camera lenses onto the photocathode of a commercial third generation image intensifier. The intensifier features a single microchannel plate (MCP)

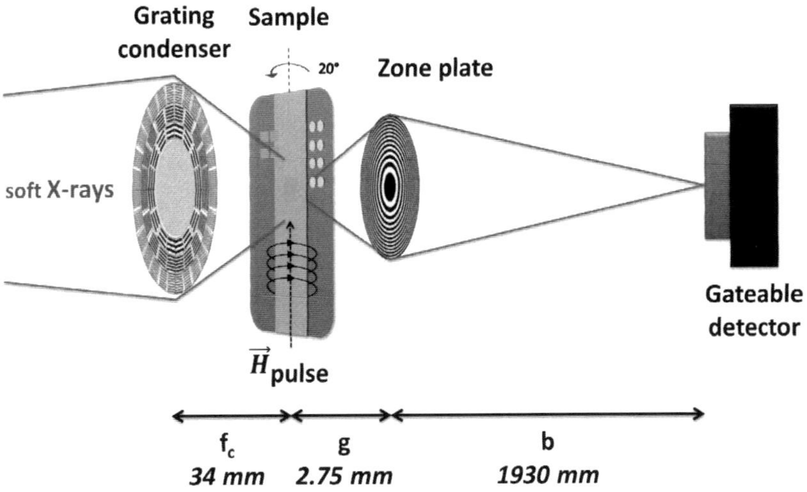

Figure 1. Setup of the TXM consisting of a grating condenser with gold central stop and a micro zone plate. Via a pulsed magnetic field $\vec{H}_{\text{pulse}}$ of a current pulse in a coplanar waveguide (figure 2) the sample can be pumped in order to follow the time evolution subsequently.

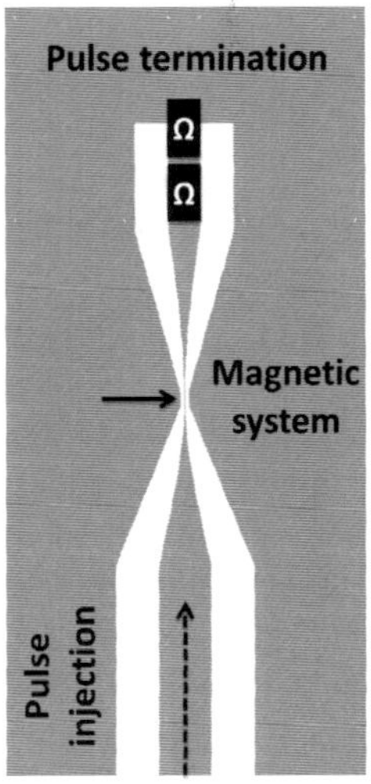

Figure 2. Geometry of the 100 Ω impedance coplanar waveguide (CPW).

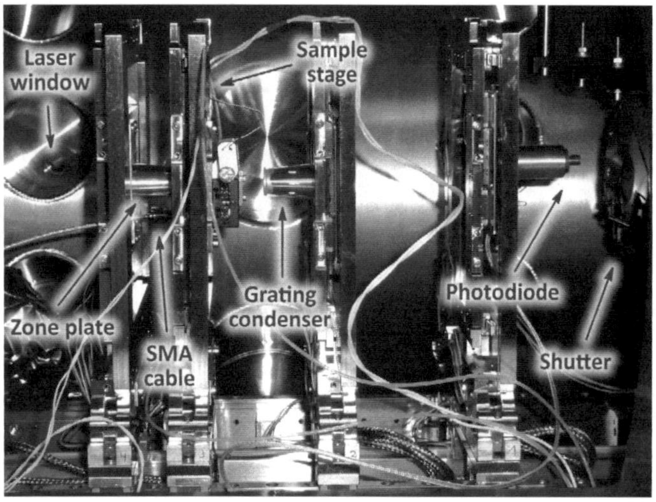

Figure 3. Photo of vacuum chamber interior with mounted components on the three-axis piezo stages. The beam enters from the right through a camera shutter. Behind the shutter a photodiode can be moved into the beam for timing purposes. The X-ray microscope consisting of condenser, sample stage and micro zone plate follows after the diode. In the back a laser coupling window is visible for excitations with a femtosecond laser.

Figure 4. Photo of the sample stage mounting the titanium sample holder with the silicon frame located on top of the water-cooled reset coil and current pulse injection circuit. An adjustable external magnet provides a bias field.

amplification stage and a P46 output screen that is fiber-coupled to a Princeton Instruments PIXIS-XF back-illuminated charge-coupled device (CCD) camera with 1024 × 1024 pixel of 13 µm size. By applying a positive bias voltage of + 30 V onto the photocathode of the image intensifier, the detector is closed by default as the generated negatively charged electrons can not leave the cathode surface. A 30 ns long negative voltage gate pulse of - 200 V amplitude synchronized to the storage ring trigger opens the detector to pass on the image information of the desired bunch containing the dynamic information by acceleration of the photoelectrons onto the MCP. The light output of the phosphor screen is collected at a fixed pump-probe delay Δt and integrated on the camera chip. The temporal overlap between pump, probe and gate pulses is achieved by moving a fast photodiode into the soft X-ray beam and monitoring all trigger signals by carefully taking into account electronic signal delays in cables and equipment.

Cobalt-platinum (Co/Pt) multilayers with perpendicular magnetic anisotropy or in-plane magnetized thin permalloy ($Ni_{80}Fe_{20}$) films of ≈ 30 nm thickness structured by electron-beam lithography and lift-off processing on a silicon substrate with a 200 nm thick silicon nitride membrane served as samples for the first measurements. Figure 2 shows the geometry of a 100 Ω impedance coplanar waveguide (CPW) [17] taking care of the current pulse transport to the nanomagnetic system and subsequent pulse termination in two microwave resistors. The conductor width is tapered down from 1 mm to 10 µm in order to increase the current density being related to the magnetic field amplitude. The CPW is directly manufactured on top of the nanostructured sample by photolithography, physical vapor deposition and lift-off processing and thus in direct contact with the magnetic layer. Proper overlap between magnetic and waveguide layer is ensured in a mask aligner by matching alignment structures written into both the magnetic layer and the waveguide lithography mask. The whole sample is tilted 70° with respect to the propagation direction of the photons to provide an XMCD sensitivity to the in-plane and out-of-plane magnetization component.

Photographs of the setup and the sample stage are provided in figure 3 and 4. Additional magnetic fields in form of an adjustable offset field generated by a permanent magnet and a pulsed reset field of up to 2 µs duration generated by a water-cooled in-vacuum coil can be applied to the sample as well.

3. Results and discussion

The TR-MTXM has been set up successfully and features down to 60 nm are observable in good agreement with the theoretically achievable resolution in this case limited by the micro zone plate optic. Figure 5 shows a raw image of Co/Pt nanodots acquired with the full-field TXM by summation of 20 camera images exposed for 250 ms and only corrected by the subtraction of a detector dark image. Because of the high degree of spatial coherence of the P04 soft X-ray beam due to the high brilliance of the PETRA III storage ring, the illumination field shows a severe speckle pattern. It was necessary to stir the beam with the refocusing Kirkpatrick-Baez (KB) mirrors of the beamline to destroy this pattern and smoothen the illumination field like done in [3]. The smoothed background can then be removed completely by correction with a flat-field image.

A first static XMCD contrast image mapping the in-plane magnetization of a 2 µm × 2 µm permalloy square is depicted in figure 6 and shows the typical Landau flux-closure pattern consisting of four equally sized domains. Black and white absorption contrast corresponds to the magnetization components pointing along or against the photon beam direction. This XMCD information becomes visible by subtracting two images taken with the circular polarized beam of positive and negative helicity. Here, for each photon helicity 40 images with an exposure time of one second were recorded at the nickel L_3 edge (852 eV).

The achieved temporal overlap of pump and probe pulses is demonstrated in figure 7. The XMCD contrast magnetization maps are constructed by subtracting the sum of six images exposed for 60 s at each photon helicity. The image sequence shows modulated permalloy nanowires of 400 nm width. Because this is a time-resolved measurement, only one of the 40 bunches in the storage ring is usable due to the limited pumping rate of the excitation source. The exposure times are longer compared to the static images due to the reduced photon flux on the camera. The spatial resolution might be worse compared to the static images in figure 5

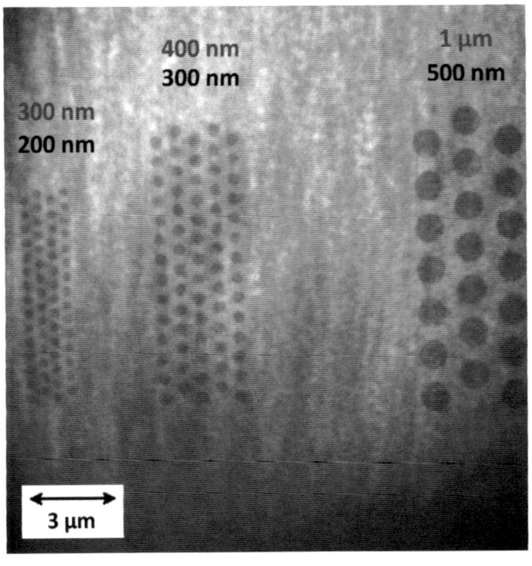

Figure 5. TXM image of nanostructured magnetic Co/Pt multilayer system with a transmission of ≈ 75 % at the cobalt L_3 edge (778 eV). The illumination field suffers from a speckle pattern due to the high degree of spatial coherence of the P04 beam that can be removed by stiring the beam. The red labels indicate the dot diameter while the black labels specify the minimum distance between two dots. This figure represents the sum of 20 camera images taken at 250 ms exposure time (dark image subtracted).

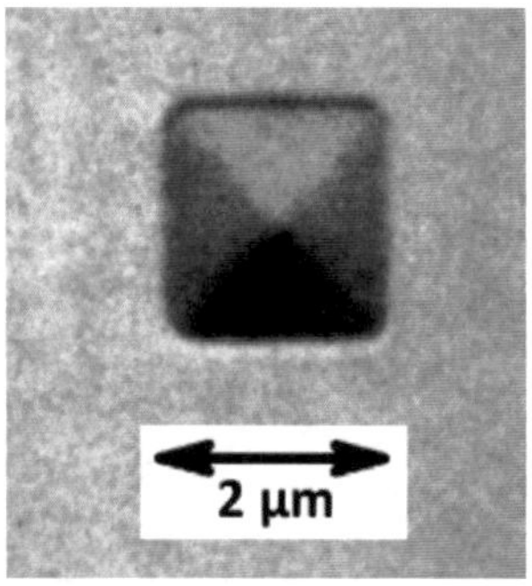

Figure 6. XMCD image of the magnetization in a 2 µm × 2 µm patterned permalloy element.

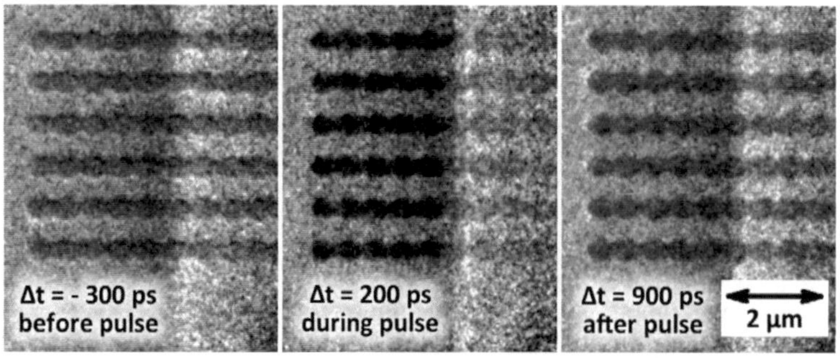

Figure 7. XMCD images of modulated permalloy nanowires of 400 nm width in an offset field before, during and after excitation by a magnetic field pulse via the CPW (visible in the left of each image).

and 6 owing to mechanical vibrations during the longer exposure times, the increased noise level due to less collected photons and because one XMCD image is generated out of the difference signal of two TXM images. Nonetheless, recent measurements show a discernibility of 40 nm features also in XMCD images (data not shown).

An offset field commonly aligns the magnetization of the nanowires in figure 7 along the field direction evident by a light contrast visible in the images before and after the magnetic pump pulse. During the presence of the pulse, however, the wire elements located in overlap with the waveguide (dark shaded area on the left) change the magnetization to the opposite direction appearing as dark contrast while the elements located next to the waveguide remain unchanged. Considering the in-plane component of the magnetic excitation pulse, this demonstrates the spatial confinement restricted mainly to the conductor width.

4. Summary and Outlook

We have set up and commissioned a new full-field TR-MTXM at the P04 beamline of PETRA III with 60 nm spatial resolution for time-resolved studies of magnetic nanostructures pumped by short magnetic field pulses and already demonstrated the time-resolved capabilities by acquiring XMCD images in temporal overlap of pump and probe pulses.

The intense magnetic field pulses producible by this setup enable a direct transient analysis of domain pattern destruction and recovery in magnetic nanostructures with possibilities to study interactions in magnetic chains and arrays. Element specific imaging of the magnetization may address questions concerning the redistribution of magnetic moments in alloys and multilayers after excitation by magnetic field pulses or femtosecond laser pulses.

A synchronized femtosecond laser system [1, 18] provides an alternative excitation source to tackle questions considering the dynamics of small magnetic systems triggered by optical pump pulses. In this case the time resolution would solely be limited by the PETRA III pulse duration of 90 ps FWHM [18] due to the already demonstrated synchronization with a residual jitter of $\sigma < 5$ ps [1, 18].

The TXM instrument will be extended to a mobile and flexible user endstation microscope with additional scanning, tomography and fluorescence imaging capabilities.

Acknowledgments

The authors would like to thank Dr. Pambos Charalambous (ZonePlates.com) for manufacturing of the grating condenser and zone plates as well as the scientific and technical staff of the beamline

P04 and the operators of PETRA III at DESY.

We gratefully thank the DFG Collaborative Research Centre 668 (SFB 668 - Magnetismus vom Einzelatom zur Nanostruktur) and the graduate school 1286 (GrK 1286 - Functional metal-semiconductor hybrid systems) for financial support as well as the German Federal Ministry of Education and Research (BMBF - Bundesministerium für Bildung und Forschung) for financial support within the Nanofocus projects (05KS7GU4 and 05KS7UL1).

References

[1] Wessels P *et al.* 2013 *J. Phys.: Conf. Ser.* **463** 012023
[2] Schütz G *et al.* 1987 *Phys. Rev. Lett.* **58** 737
[3] Vogt U, Lindblom M, Charalambous P, Kaulich B and Wilhein T 2006 *Opt. Lett.* **31** 1465
[4] Romming N *et al.* 2013 *Science* **341** 636
[5] Parkin S S P, Hayashi M and Thomas L 2008 *Science* **320** 190
[6] Cowburn R 2002 *J. Magn. Magn. Mater.* **242 - 245, Part 1** 505
[7] Loth S, Baumann S, Lutz C P, Eigler D M and Heinrich A J 2012 *Science* **335** 196
[8] Beaurepaire E, Merle J C, Daunois A and Bigot J Y 1996 *Phys. Rev. Lett.* **76** 4250
[9] Pfau B *et al.* 2012 *Nat. Commun.* **3** 1100
[10] Stanciu C D *et al.* 2007 *Phys. Rev. Lett.* **99** 047601
[11] Kirilyuk A, Kimel A V and Rasing T 2010 *Rev. Mod. Phys.* **82** 2731
[12] Uhlíř *et al.* 2013 *Nat. Nanotechnol.* **8** 341
[13] Jung H *et al.* 2010 *Appl. Phys. Lett.* **97** 222502
[14] Vogel A *et al.* 2011 *Phys. Rev. Lett.* **106**(13) 137201
[15] Stöhr J and Siegmann H C 2006 *Magnetism: From Fundamentals to Nanoscale Dynamics* (Berlin: Springer)
[16] Fischer P 2011 *Mater. Sci. Eng. R Rep.* **72** 81
[17] Wen C 1969 *IEEE T. Microw. Theory* **17** 1087
[18] Schlie M 2013 *Time-resolved Studies at PETRA III with a Highly Repetitive Synchronized Laser System* Ph.D. thesis University of Hamburg

Journal of Physics: Conference Series **499** (2014) 012010 doi:10.1088/1742-6596/499/1/012010

Design of sample holder for *in operando* XAS of oxidation state changes in $La_{1-x}Sr_xMnO_{3\pm\delta}$ electrodes

M L Traulsen

Technical University of Denmark, Department of Energy Conversion and Storage
Frederiksborgvej 399, 4000 Roskilde, Denmark

E-mail: matr@dtu.dk

Abstract. The design and construction of a sample holder for *in operando* X-ray absorption spectroscopy on solid oxide electrodes, like for instance the lanthanum strontium manganite electrodes, is explained. The sample holder is mounted in a test-cell, which is also described together with some preliminary test-results from temperature measurements and electrochemical characterisation performed on a "dummy" sample in the test cell. The test-cell is a modified version of a test-cell previously designed in our laboratory, and will in the future be used for X-ray absorption experiments on solid oxide electrodes.

1. Background

The perovskite $La_{1-x}Sr_xMnO_{3\pm\delta}$ (LSM) has found application as electrode material in solid oxide electrochemical cells, like the solid oxide fuel cells (SOFCs), solid oxide electrolysis cells (SOECs) and cells for electrochemical gas purification. During several decades extensive research has been conducted on LSM, including work on electrical properties [1], defect chemistry [2] and the influence of electrical polarisation on segregation in LSM electrodes [3] An overall tendency today in the research on LSM electrodes and solid oxide electrodes in general, is an increased interest in studying the electrodes at realistic operating conditions, i. e. *in operando*, or at conditions close-to realistic operating conditions, i.e. *in situ,* with optical/spectroscopic methods. For *in operando/in situ* studies the use of X-ray absorption techniques has great advantages, as they in contrast to many other analytical methods do not require vacuum conditions. The challenge in performing *in operando/in situ* studies of solid oxide electrodes is the need for operating the electrodes at elevated temperatures (300-900 °C), in controlled atmosphere and with electric contact, while still maintaining the optical access to the electrodes. To meet these requirements a careful design is necessary for any test-cell and sample holder for *in operando* studies of solid oxide electrodes. The design of a test-cell for *in operando* X-ray absorption spectroscopy of solid oxide electrodes has been described by Hagen et al. together with preliminary test-results [4]. This paper describes the design of a new sample holder for a test-cell, which is a modified version of the test-cell described by Hagen et al. [4].

2. The design of the test-cell.

The test-cell is intended for measurements in fluorescence mode, as both sample thickness and the mounting of the sample on a heat-stage make measurements in transmission mode impossible. In order to make it easy to transport the cell between different synchrotron facilities, the test-cell has a fairly compact design, with the dimensions of the main house being 8 cm x 12 cm x 12 cm. In the following the main house of the test-cell, the sample-holder and the sample, which in the future will be measured in the cell, will be described in more detail.

2.1 The main-house

The main house is made of aluminium and has a removable front plate and back plate for easy access to the sample in the inner of the house. On the two opposite sides of the test-cell are X-ray windows, and a gas inlet and a gas-outlet made with standard fittings. The front plate is dominated by a large window, 9 cm in diameter, to allow for detection of the fluorescence signal from the sample. The small X-ray windows on the sides of the house and the big X-ray window in the front is made of Kapton® foil which is clamped between the front-plate/metal rings and the main-house together with O-rings to secure the sealing. Figure 1 shows a schematic drawing of the cell design, and a front view and a side view image of the cell.

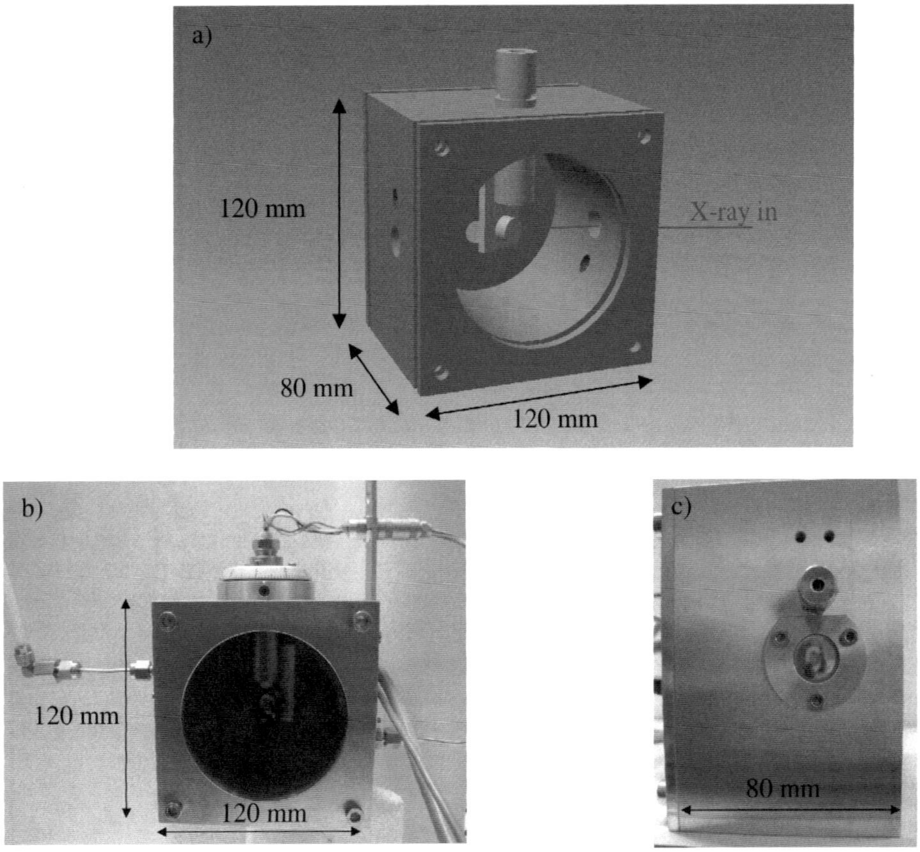

Figure 1. Illustrations of the test-cell for *in operando* X-ray absorption spectroscopy on solid oxide electrodes, a) schematic drawing, b) front view image and c) side view image.

Journal of Physics: Conference Series **499** (2014) 012010 doi:10.1088/1742-6596/499/1/012010

2.2 The sample holder

The sample holder is a hollow stainless steel rod, on which a heat-stage is mounted. The stainless steel rod is mounted through "the roof" of the test-cell, where an O-ring between rod and house secure the sealing. The steel rod is turnable, meaning the angle between the incoming X-ray and the sample may be adjusted, and the actual angle may be read of on a scale ring on top of the house. In the bottom of the steel rod, inside the main house, the heat-stage is mounted. Rather than mounting the heat-stage directly on the steel rod, the heat-stage is mounted in an alumina plate mounted in the rod. This design is made to lower the heat transfer from the heat-stage to the steel rod. The heat-stage (Heatwavelabs) is made of alumina and has a diameter of 0.5". The sample is mounted on the heat-stage with an alumina ring and to small clamps and Au mesh is used as current collector (CC) on the front side of the sample while painted Au is used for current collection on the back side.

Through the hollow middle of the steel rod is mounted a 6-hole alumina tube. This tube guides 4 leads for electrochemical characterization and 2 leads for a thermocouple to the sample on the heat-stage. The leads for electrochemical characterization are connected to a Gamry Ref600 potentiostat, thus making electrochemical characterization and electrical polarization of the sample possible. On Figure 2 is shown a close-up of the sample holder.

Figure 2. Close-up of sample mounted on sample holder. The sample is mounted below the Au mesh current collector, which is fastened to the sample by two clamps and an alumina ring.

2.3. The sample

Thin oxide films (100-1000 nm thick) are widely used as model electrodes, to study the electrochemical processes occurring in solid oxide electrodes [5]. Similarly, in the test-cell described here, thin film electrodes will be applied to monitor the effect of the electrical polarisation on the mangane oxidation state in the LSM electrodes. According to theory cathodic polarisation of LSM electrodes will cause an increase in the number of oxygen vacancies, accompanied by a decrease in average oxidation state of the manganese to retain charge neutrality [6]. Orikasa et al. [7] very briefly reported experiments, confirming this theory, but as the observed effect was close to the uncertainty of the experiments, a repetition of the experiments and confirmation of the results is desirable.

In the test-cell described here, the sample will be an electrochemical cell, (10 mm x10 mm x 0.5 mm) with 3 layers: one thin-film LSM electrode, an YSZ single crystal electrolyte, and on the opposite side of the electrolyte a porous LSM-CGO composite electrode. The porous composite electrode is used, as this electrode has a significantly lower polarisation resistance compared to the thin-film electrode. This means electrochemical impedance spectra recorded on the electrochemical cell will be entirely dominated by the thin film electrode, and thus changes in thin film electrode induced by the electrical polarisation will be detectable both on the electrochemical impedance spectra and in the X-ray absorption spectra.

Journal of Physics: Conference Series **499** (2014) 012010 doi:10.1088/1742-6596/499/1/012010

3. Preliminary measurements in the test cell

So far, the described test cell has not as intended been used for X-ray absorption experiments on thin film LSM electrodes. However, preliminary tests of the cell's capability to control the sample temperature has been made, and the feasibility of performing electrochemical measurements in the cell has been investigated, in both cases using a "dummy sample" consisting of a YSZ single crystal with painted Pt electrodes on side. The results are reported in the following two sections.

3.1. Temperature control and measurement

There are three different ways to evaluate the temperature in and near the sample: a) By the thermocouple inside the heat-stage, this thermocouple is used to control the heat-stage temperature, b) By the thermocouple mounted near the sample, or c) From the ohmic resistance of the YSZ single crystal electrolyte of the sample. The latter is easily measured by electrochemical impedance spectroscopy, and from the YSZ ohmic resistance the temperature may be calculated from the correlations found by for instance Cheikh et al. [8]. It is believed using the ohmic resistance is the most accurate way to measure the sample temperature, as the method determines the temperature in the sample. In contrast, the thermocouple measurements may be inaccurate as the thermocouples either not are contacting the sample directly or by contacting the sample might affect the actual sample temperature. In Figure 3 the temperature of a dummy sample determined from the ohmic resistance is plotted as a function of the heat-stage temperature measured by a thermocouple. The difference between the heat stage temperature and the sample temperature increases from 40 °C to 125 °C in the range 400 °C to 700 °C, which is ascribed to contacting problems between sample and heat stage and heat loss from the sample to the surroundings.

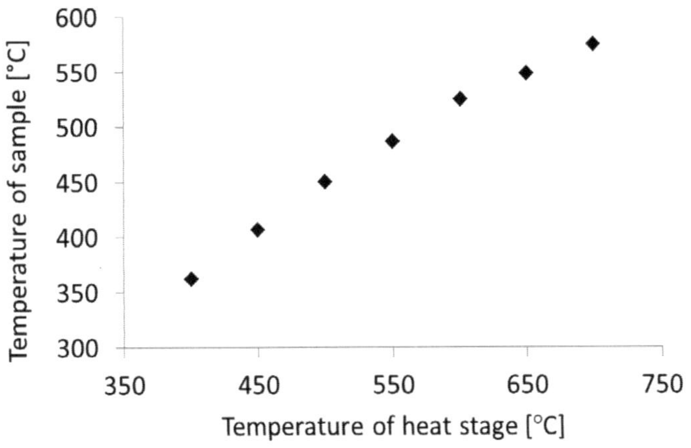

Figure 3. Temperature of a sample as function of the heat stage temperature. The sample temperature was determined from the ohmic resistance of the YSZ electrolyte, while the heat stage temperature was measured by a thermocouple mounted inside the heat stage.

3.2. Electrochemical test on dummy sample

To check the feasibility of performing electrochemical measurements in the test cell, electrochemical impedance spectroscopy was conducted on the dummy sample consisting of a YSZ single crystal with a Pt electrode on each side. As seen in Figure 4, a nice impedance spectrum was obtained, confirming the test cell can be used for electrochemical characterisation.

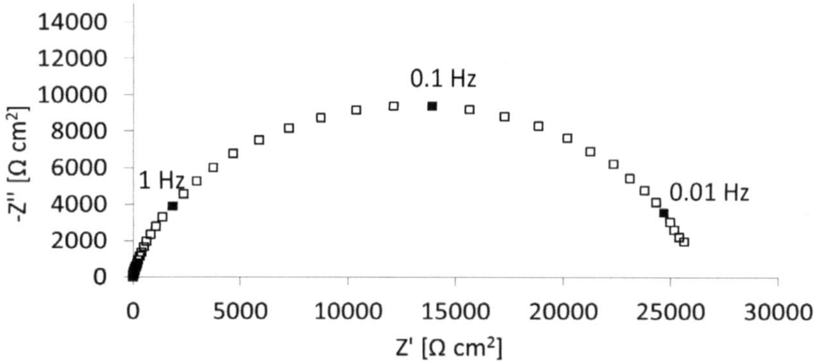

Figure 4. Electrochemical impedance spectrum recorded in the test cell on an electrochemical cell with a YSZ electrolyte and two Pt electrodes. The sample was at 525 °C in a gas flow of 50 ml/min 4% O_2 in Ar.

4. Outlook

The construction of a sample holder for *in operando* XAS on solid oxide electrodes has been described together with a description of the applied test-cell. It has been confirmed the cell is capable of controlling the sample temperature and preliminary electrochemical measurements has been made. The test-cell is a modified version of the cell previously used and described by Hagen et al. [4]. In the near future the cell will be used for *in operando* experiments on LSM thin film electrodes. The outcome of the experiments will hopefully be experimental confirmation of a decrease in the average manganese oxidation state during cathodic polarisation of LSM electrodes, as predicted by theory and indicated by the experiments by Orikasa et al. [7].

References

[1] J. H. Kuo, H. U. Anderson, and D. M. Sparlin, J. Solid State Chem. 87 (1990) 55.

[2] J. Mizusaki, N. Mori, H. Takai, *et al*, Solid State Ionics 129 (2000) 163.

[3] M. Backhaus-Ricoult, K. Adib, T. S. Clair, B. Luerssen, L. Gregoratti, and A. Barinov, Solid State Ionics 179 (2008) 891.

[4] A. Hagen, M. L. Traulsen, W. R. Kiebach, and B. S. Johansen, J. Synchroton Rad. 19 (2012) 400.

[5] T. Ioroi, T. Hara, Y. Uchimoto, Z. Ogumi, and Z. Takehara, J. Electrochem. Soc. 145 (1998) 1999.

[6] A. Hammouche, E. Siebert, A. Hammou, M. Kleitz, and A. Caneiro, J. Electrochem. Soc. 138 (1991) 1212.

[7] Y. Orikasa, T. Ina, T. Nakao, T. Fukutsuka, A. Mineshige, K. Amezawa, T. Kawada, and Y. Uchimoto, Meet. Abstr. - Electrochem. Soc. 1001 (2010) 531.

[8] A. Cheikh, A. Madani, A. Touati, H. Boussetta, and C. Monty, Journal of the European Ceramic Society 21 (2001) 1837.

Journal of Physics: Conference Series **499** (2014) 012011 doi:10.1088/1742-6596/499/1/012011

Palm-top size X-ray microanalyzer using a pyroelectric focused electron beam with 100-micro-meter diameter

Jun Kawai, Issei Ohtani, Akira Imanishi, Susumu Imashuku

Department of Materials Science and Engineering, Kyoto University, Sakyo-ku, Kyoto 606-8501, Japan

E-mail: kawai.jun.3x@kyoto-u.ac.jp

Abstract. We have developed a palm-top size EPMA (electron probe X-ray microanalyzer), operated by 3 V electric battery except for a rotary vacuum pump. The electron beam was generated by a pyroelectric single crystal, $LiTaO_3$. A needle was used to make a focused electron beam. The smallest beam size was 100 μm on the sample surface. The X-ray spectra were measured through a Kapton window by a Si-PIN detector for a model specimen containing TiO_2 and MnO_2 particles, which was an aerosol model specimen, where TiO_2 and MnO_2 particles of size about 100-200 μm were separated by a few hundreds micrometers. By moving the sample stage manually, the X-ray spectra were measured for 300 s each by 300 μm e-beam, and the measured X-ray intensities were strong enough for identification of the major element in individual 100-200 μm size aerosol particles.

1. Introduction

Brownridge found X-ray emission effect from a pyroelectric single crystal when it was placed in a vacuum and it was heated or cooled [1]. The conditions of the pyroelectric X-ray emission were extensively studied by Brownridge and Reboy [2] and by Brownridge [3]. An X-ray source using the principle found by Brounridge is commercially available from Amptek Ink. as COOL-X [4]. Applications of COOL-X for X-ray fluorescence analysis and even X-ray absorption spectroscopy were summarized by Kawai *et al.* [5]. Kawai *et al.*[6] also published methods how to make such an X-ray source using a cheap and easily obtainable electric, mechanical, and vacuum parts in a chapter of an open access book.

Geuther and Sanon [7] showed a series connection of two pyroelectric crystals in an electric circuit produced twice the higher energy X-rays; one crystal produced 100 keV X-rays, and two crystals could produce X-rays up to 200 keV. Hiro *et al.* [8] put a small specimen piece on the top of one pyroelectric single crystal ($LiTaO_3$ of 3 × 3 mm^2 cross-section and 10 mm length) during a similar experiment as Geuther and Danon, and observed strong characteristic X-rays from the small piece specimen. This was the first experiment of the "palmtop EPMA". The palmtop EPMA has been improved step by step by using quick coupling nipples and T-shape pipes [9-11]. Carbon rod sample holder could avoid the X-rays from the sample stage. And finally by using a needle on the top of the pytoelectric crystal, we have succeeded to make an electron beam of the size of 100 μm diameter [12].

When the temperature of the pyroelectric single crystal increased from room temperature to *e.g.* 100 °C by a small Peltier device, the top surface of the pyroelectric crystal accumulated electric charge and consequently the electric potential becomes about -40 kV. A floating residual electron in the vacuum was accelerated toward to the grounded sample on the carbon rod, hitting the sample surface.

Once the sample was bombarded by the electron, then the secondary electrons will be produced, and again the secondary electrons were accelerated by the electric potential to hit the target again and again, exciting the characteristic X-rays. The X-rays continued for a few minutes and stopped by neutralized the electric charge on the ptroelectric crystal. Then the temperature of the pyrpelectric crystal should be decreased down to room temperature by changing the polarity of the 3 V electric battery of the Peltier device. Then again the temperature was increased, we could repeat the X-ray emission. If a needle was put on the top surface of the pyroelectric single crystal, and high electric potential surfaces other than the needle was shielded by vacuum silicone grease, the electron beam could be focused to 100 μm on the sample surface. In the present paper, we describe how to find the focusing of the electron beam during the experiment, and present an example of practical single particle model specimen analyzed by a 300-μm electron beam excited X-ray emission, where the X-ray intensity was 9 times stronger than 100-μm diameter e-beam.

2. Experimental

A photo of the pyroelectric crystal used for the palmtop EPMA (electron probe X-ray microanalyzer) was shown in Figure 3 of Ref.[10], a previous proceedings paper of this series of international conference (ICXOM 2011). As a first experiment after ICXOM2011, we installed a metal needle (Figure 1) on the top of the pyroelectric single crystal. However, the electron beam was not focused, and was radiated in all directions observed by a fluorescent screen and a CCD camera [13]. This experiment was set in a dark vacuum chamber evacuated by a rotary pump (1 Pa).

Then the next step of experiment, we buttered vacuum silicone grease on the surface of the pyroelectric crystal and the metal base (Figure 1) of the needle, just like buttering a bread. Since the vacuum was a low vacuum, evacuated by a rotary pump (1 Pa), and the vacuum chamber was made of quick couplings, this buttering was quite easily finished within a few minutes. Then we were surprised that the electron beam was focused on the fluorescent screen as a tiny spot as shown in Figure 2. The place of the beam spot on the fluorescent screen was always the nearest place from the tip of the needle.

Figure 1. Photo of a needle. The base is metallic cube (3 × 3 × 3 cm^3). This photo was taken after the vacuum silicone grease was buttered. The needle length is 16 mm. The smallest scale is 1 mm.

We have tried to make much smaller the spot size, and finally the size became 100 μm diameter [12], by reducing the emission current, changing the distance between the target and the tip, length of the needle, and kind of metal (Fe, Cu, W or Au) and sharpness of the needle. The emission current could be controlled by changing the temperature gradient of the pyroelectric crystal. The emission current was most effective for the beam size. We are now making an effort to make much smaller the beam size.

The beam size was determined by the brightness read from the CCD data, as shown in Figure 3. The measured brightness was plotted, and was fitted by a Gaussian function. The full width at half maximum (FWHM) of the Gaussian function was used as the electron beam size.

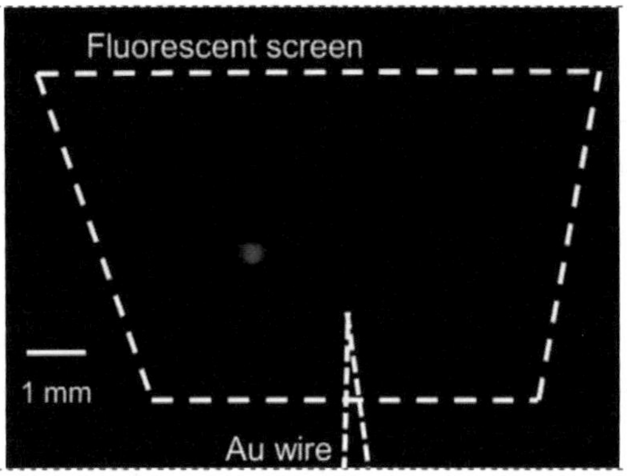

Figure 2. The beam spot observed by a fluorescent screen and by a CCD camera.

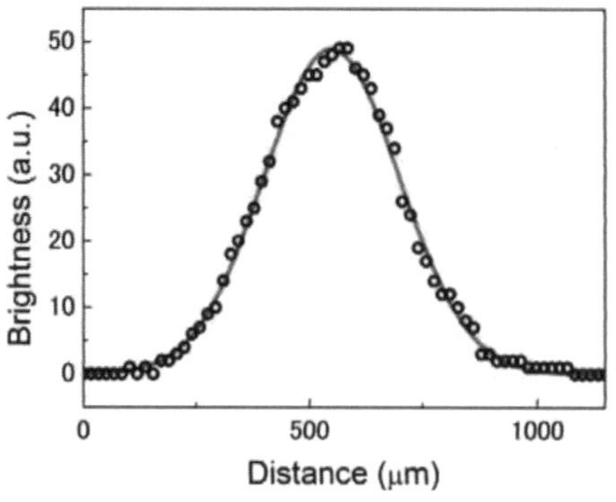

Figure 3. The recorded beam spot brightness of Figure 2, and Gaussian fitted curve. The FWHM is about 300 μm.

3. Results and discussion

The practical size of the electron beam at present is now 300 μm. However even such a relatively large spot size compared with genuine EPMA or SEM (scanning electron microscope), we can demonstrate some usage of our palmtop EPMA. Figure 4 shows a model specimen we tried. TiO_2 and MnO_2 particles of size about 100-200 μm were put on a double-sided adhesive carbon tape, but separated about a few hundreds micrometers as shown in Figure 4, which is a SEM image (JEOL JSM-5610LVS, Tokyo) of the model specimen. We placed TiO_2 particle at the beam position of our palm-top EPMA in order to measure the X-ray spectrum, and then placed the MnO_2 particle, by moving the sample stage manually. The X-ray spectra were measured through the Kapton window for 300 s each by using a Si-PIN X-ray detector (Figures 5 and 6). The X-ray intensities were strong enough for identification of the major element in 100-200 μm size particles. The present model specimen can be regarded as aerosol particles collected by an impactor though still the particle sizes were still too large.

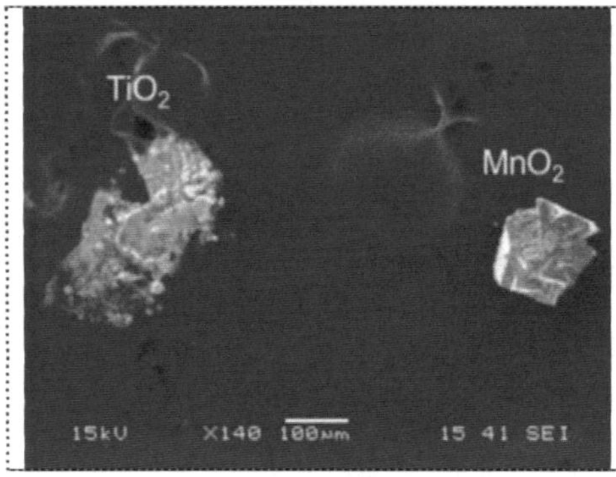

Figure 4. A secondary electron image of an SEM for two different particles in order to demonstrate the microarea analysis by the present palmtop EPMA.

By comparing Figures 5 and 6, we can find that the Ti Kα is about 60 counts while Mn Kα is 400 counts. The particle size of TiO_2 in Figure 4 was about twice as large as that of MnO_2. Therefore the electron beam when TiO_2 was measured was a little bit away from the centre of the particle. The MnO_2 spectrum was strong enough and thus the continuum X-ray spectrum was also observable. The maximum intensity of the continuum X-rays was about 4 keV, and thus the effective electron beam acceleration voltage to excite the sample X-rays was quite similar to those of the conventional EPMA or SEM-EDX, *i.e.* 15-25 kV acceleration voltage. The duration time was 300 s for each of the sample

Journal of Physics: Conference Series **499** (2014) 012011 doi:10.1088/1742-6596/499/1/012011

particle, and thus the heating and cooling cycle of the Peltier device was more than one cycle to measure one X-ray spectrum.

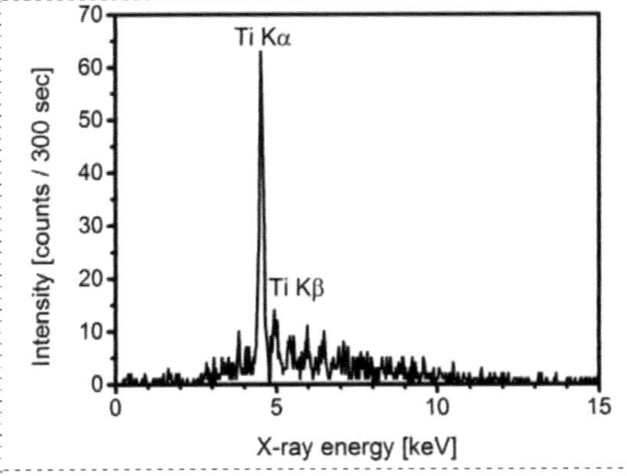

Figure 5. Measured X-ray spectrum at the beam position of TiO_2 particle in Figure 4.

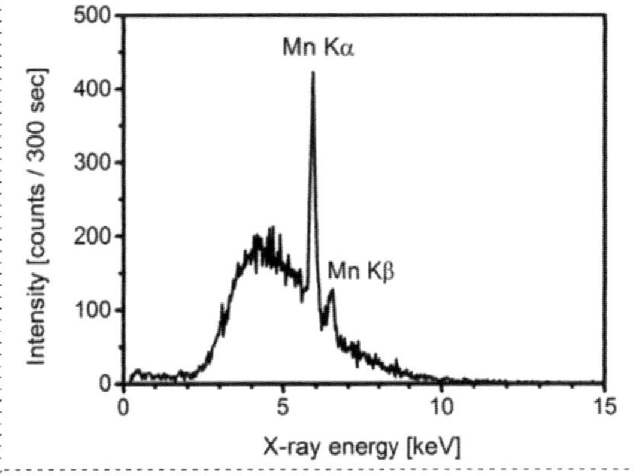

Figure 6. Measured X-ray spectrum at the beam position of MnO_2 particle in Figure 4.

4. Conclusions

We have described how easily we discovered a method of focusing an electron beam using a pyroelectric single crystal and a needle. The buttering of vacuum grease was the key but it was quite easily achieved because we have used quick-coupling vacuum components. We also demonstrated that even a palmtop size EPMA operated by 3 V electric battery except for a rotary vacuum pump, could measure X-ray spectra of a few hundreds μm size single particle within 5 minutes, but the intensities of which were from 60 to 400 counts for 5 minutes. The electron beam was generated by a metal needle buttered by vacuum silicone grease attached to a pyroelectric $LiTaO_3$ single crystal. The smallest beam size was 100 μm on the sample surface at present, but practically 300 μm diameter was easy to produce and the X-ray intensity was not too weak.

Acknowledgement

The present study was financially supported by Asahi Glass Foundation.

References

[1] Brownridge J D 1992 *Nature* **35**8 287
[2] Brownridge J D and Reboy S J 1999 *Appl. Phys.* **86** 640
[3] Brownridge J D 2004 Appl. Phys. Lett. **85** 1298 (2004).
[4] www.amptek.com
[5] Kawai J, Ishii H, Ida H 2012 *X-Ray Spectrom.* **41** 216
[6] Kawai J, Nakaye Y and Imashuku S 2012 *The Scanning Electron Microscope* ed V Kazmiruk, (Rijeka: InTech) pp 89-100
[7] Geuther J A and Danon Y 2005 *J. Appl. Phys.* **97** 104916
[8] Hiro E, Yamamoto T and Kawai J 2010 *Adv. X-Ray Chem. Anal. Japan* **41** 195
[9] Imashuku S, Imanishi A and Kawai J 2011 *Anal. Chem.* **83** 8363
[10] Imashuku S, Imanishi A and Kawai J 2012 *Proceedings of the 21st International Congress on X-Ray Optics and Microanalysis (Campinas, Brazil, 5–9 September 2011) (AIP Conference Proceedings* vol 1437*)* eds C A Pérez and A M de Souza (New York: AIP) pp 29-31
[11] Imashuku S and Kawai J 2012 *Rev. Sci. Instrum.* **83** 016106
[12] Imashuku S, Imanishi A and Kawai J 2013 *Rev. Sci. Instrum.* **84** 073111
[13] Fuyuno N, Imashuku S and Kawai J 2102 *J. Spectrosc. Soc. Japan* **61** 190

Journal of Physics: Conference Series **499** (2014) 012012 doi:10.1088/1742-6596/499/1/012012

Retrieving past geodynamic events by unlocking rock archives with μ-XRF and μ-spectroscopy

V De Andrade[1,2], J Ganne[3], B Dubacq[4,5], C G Ryan[6], F Bourdelle[7], A Plunder[4,5,8], G Falkenberg[9], J Thieme[2]

[1]Advanced Photon Source, Argonne National Laboratory, Argonne, IL 60439, USA
[2]Photon Science, Brookhaven National Laboratory, NSLS-II, Upton, NY 11793, USA
[3]IRD, UR 234, GET, Université Toulouse III, 31400 Toulouse, France
[4]UPMC Univ. Paris 06, ISTEP, UMR 7193; F-75005 Paris, France
[5]CNRS, ISTEP, UMR 7193; F-75005 Paris, France
[6]CSIRO Earth Sciences & Resource Engineering, Normanby Rd, Clayton VIC 3168, Australia
[7]GeoRessources, UMR 7359, Université de Lorraine, 54506 Vandoeuvre-lès-Nancy, France
[8]UMR CNRS 8538 Laboratoire de Géologie, ENS, 24 rue Lhomond, F-75005 Paris, France
[9]Photon Science, Deutsches Elektronen-Synchrotron DESY, Notkestr. 85, 22603 Hamburg, Germany

Email: vdeandrade@anl.gov

Abstract. Rocks are commonly polycrystalline systems presenting multi-scale chemical and structural heterogeneities inherited from crystallization processes or successive metamorphic events. This work illustrates how spatially resolved analytical techniques coupled with thermodynamic approaches allow rock compositional variations to be related to large-scale geodynamic processes. Emphasis is placed on the contribution of quantitative chemical imaging to the study of 2.2–2.0 Gy old metamorphic rocks from the West African Craton. A thorough analysis of elemental chemical maps acquired on rock thin sections enabled high pressure relic minerals to be located and re-analyzed later with precise point analyses. The pressure-temperature conditions of crystallization calculated from these analyses are typical of modern subduction zones. These results push back the onset of modern-style plate tectonics to 2.15 Gy, i.e. more than one billion years earlier than was consensually accepted. The second part of the paper describes the imaging capabilities offered by the new ultra-bright diffraction limited synchrotron sources. Experimental data acquired with the Maia detector at beamline P06 at Petra III as well as simulations of μ-XRF spectra that will be generated at the SRX beamline at NSLS-II are presented. These results demonstrate that cm^2 large chemical maps can be acquired with submicron spatial resolution and a precision suitable for thermobarometric estimates, with dwell time smaller than 1 millisecond. The last part of the paper discusses the relevance of utilizing recent X-ray fluorescence nanoprobes for diagenetic to low grade metamorphism applications.

1. Introduction

Rocks are products and witnesses of geodynamic events shaping the Earth and planetary bodies. A rock is defined by its mineralogy that is inherited from its genesis and from the successive events it encountered during its lifetime. Historically, geologists endeavor to characterize the mineralogy of rocks, from which they infer large-scale processes, for example derived from plate tectonics. With the constant improvement of analytical tools in terms of sensitivity and spatial resolution, the apparent degree of complexity of rocks keeps increasing. The amount of information stored in rock-forming minerals is indeed extremely large, and more information about these minerals leads to a better understanding and more accurate modeling of rock-forming processes. These processes resulted in heterogeneities in the physical and chemical properties of continental crustal rocks over scales varying from the geological unit (10^2-10^3 meters) to the mineral (10^{-3}-10^{-4} meters) ; several studies within the past 10 years focused on heterogeneities occurring at the micrometer and nanometer level[1-5]. Thanks to elemental chemical imaging acquired either with electron probe micro-analyzer (EPMA), transmission electron microscope (TEM) or synchrotron X-ray fluorescence microprobe, scientists were able to characterize micrometric subdomains and the modalities of their interconnections. Combining thermodynamic models with chemical maps, it is possible to calculate the conditions of crystallization of the rock-forming minerals, which are combined to obtain high resolution pressure-temperature (P-T) paths in favorable samples[1-3, 6]. Such P-T paths are of primary importance for the understanding of geological processes. If EPMA remains one of the main analytical tool of geologists, it shows important limitations when it comes to chemical imaging, analysis of trace elements and spatial resolution below a few microns. A typical acquisition time for a single analysis of major elements is about 1 minute. Such a dwell time is incompatible with chemical imaging, where hundreds of thousands - if not millions - of analyses are required. Ultra-high brightness synchrotron X-ray fluorescence microprobes are currently the most suitable instruments to overcome the physical limitations of non-synchrotron laboratory techniques. The current light source construction projects like NSLS-II, MAX-IV or the upgrades of ESRF and APS[7] aiming for ultra-low emittance sources are about to enable very fast imaging of elements in concentration down to the sub-ppm level. The several-orders-of-magnitude gain in flux in comparison with current facilities will be also highly favorable to 3D imaging, boosting time consuming tomography techniques with a micro or nano pencil beam in combined fluorescence and diffraction mode.

The first part of this paper illustrates the contribution of micro X-ray fluorescence (μ-XRF) imaging and microspectroscopy to the study of metamorphic rocks. The second part highlights the possibilities offered by intense new generation synchrotron fluorescence microprobes like the Submicron Resolution X-ray Spectroscopy beamline[8, 9](SRX) at the National Synchrotron Light Source II (NSLS-II). Recent detector developments are also discussed. The third section focusses on the interest of zooming at the nanoscale with XRF nanoprobes while the last one concludes on radiation damage that may occur with high fluence X-ray beams.

2. Contribution of fluorescence microprobes to the study of the 2.2-2.0 Gy old West African metamorphic province

This part comments on a study[10] performed on the West African Craton (WAC). As most of the other metamorphic units in Archaean and Palaeoproterozoic cratons, WAC mainly consists of volcano-sedimentary and volcano-plutonic basins metamorphosed in greenschist facies. WAC rocks are principally composed of phyllosilicate called chlorite (chl) and phengite (phg) stable over large ranges of pressure and temperature. Previously, the lack of detailed thermodynamic data on these mineralogical phases has been a hindrance to obtaining precise and accurate P-T estimates of greenschist facies rocks. Owing to recent advances in our knowledge of the thermodynamic properties of these minerals[11-14], detailed investigations of the thermal evolution of the WAC has been possible.

Hundreds of rock thin sections were prepared from samples collected during field missions in Burkina Faso. Following visible light microscopy, tens of samples were selected to be analyzed with point EPMA analyses, while EPMA images were acquired on a few of the most promising samples. Chl and phg were preferentially analyzed. From the measured compositions, the thermodynamic multi-equilibria technique[15] can be applied. This approach checks if different mineralogical phases are in thermodynamic equilibrium and returns the P-T conditions of equilibrium, interpreted as conditions of crystallization. Below ~550°C, it has been demonstrated that rocks do not fully re-

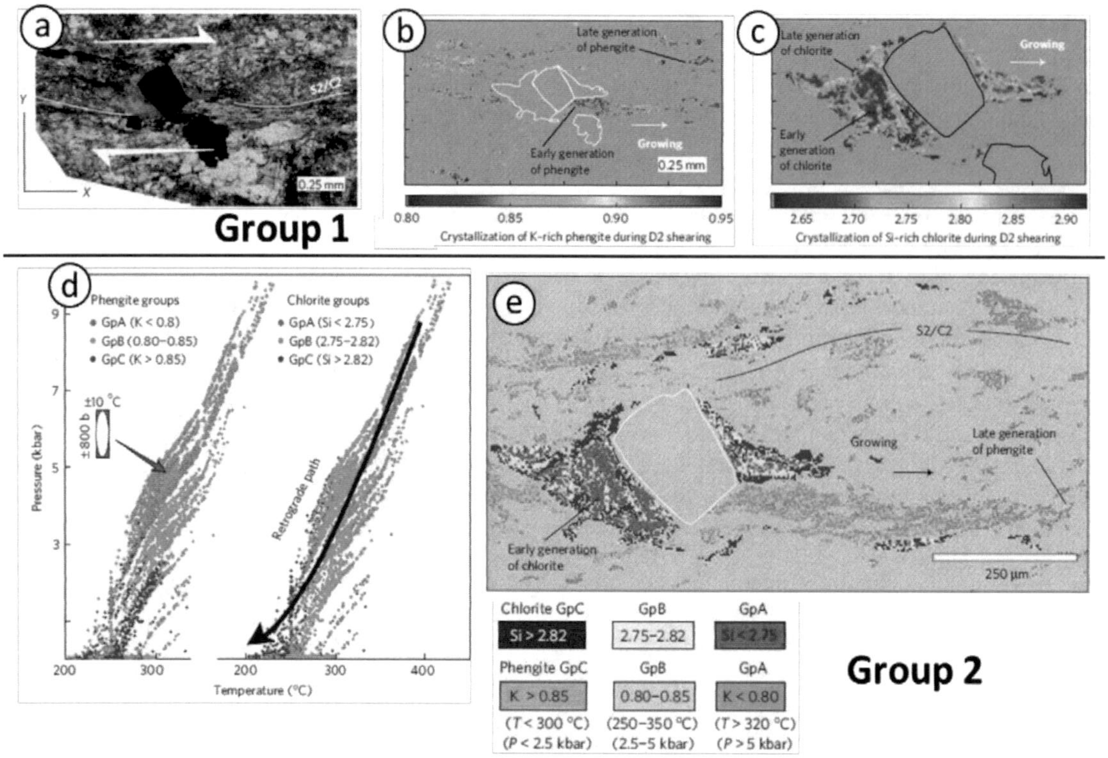

Figure 1. Group 1: Chemical evolution of chl and phg during deformation in sample Fa-33. **(a)** Photomicrograph of sample Fa-33. **(b)** Distribution of K-rich phg around a crystallization tail. **(c)** Distribution of Si-rich chl around haematite grains. Color bars correspond to the number of K & Si cations in phg **(b)** and chl **(c)**. **Group 2:** P-T calculations based on chl-phg analyses. **(d)** P-T path vs K-content (left) and Si-content (right) in phg & chl. The P-T path was calculated from conventional spot analysis. **(e)** P-T maps derived from compositional images. Each chl and phg group is characterized by a specific range of composition, P & T determined in **(d)**. The Si-poor chl and K-depleted phg corresponding to the high P group A are mostly observed in the early-formed microstructures *(modified after Ganne et al., 2011)[10]*.

equilibrate. Dissolution / crystallization processes take over intra-crystalline diffusion. New crystallizations occurred locally while environmental constraints like P & T evolve. Metamorphic rocks end up as a mosaic of subdomains of different compositions whose spatial organization is a result of deformation structures, rock protolith, fluid chemistry and rock trajectory in the continental crust.

In parallel to EPMA analyses that are insensitive to the speciation of elements, µ-XANES spectra were acquired on the ID21 beamline at the ESRF to quantify the $Fe^{3+}/(Fe^{2+}+Fe^{3+})$ ratio in several chl and phg grains. The amount of Fe^{3+} incorporated in phyllosilicate varies significantly and has important consequences for the P-T calculations. A thermodynamic method allowing the inference of this ratio in chl has been proposed[2] and validated at the ESRF on the ID24 beamline, with a µ-XANES imaging approach[16]. This method has been extended here to phg and confirmed with µ-XANES spectra acquired on ID21 as control points. In some phg, the amount of Fe^{3+} is greater than 80%. Building on these new methods and considering in addition the interlayer water content in phg[11], the multi-equilibria technique was successfully applied. For most of the samples, more than a hundred P-T conditions of crystallization were returned, defining high resolution P-T paths (fig. 1d). The most remarkable results came from two samples, where the calculated pressure of crystallization reached 10-12 kb for temperatures lower than 450 °C, and where a segment of their P-T path follows a cold thermal gradient around 10-12 $°C.km^{-1}$. These conditions are typical of modern subduction zones, suggesting that modern-style plate tectonics already existed during the Palaeoproterozoic era. These results push back the onset of modern-style plate tectonic to 2.15 Gyr, i.e. more than one billion years earlier than what was consensually accepted.

P-T path calculation of rocks from the WAC has been relatively straightforward. However, finding the relic minerals present in minute quantities and that crystallized at high P in a subduction zone has proved challenging. Amongst several hundred phg analyses, only one was in thermodynamic equilibrium with the tested chl, a result statistically not robust enough to question the date of the onset of modern-style tectonics. If the μ-XRF images obtained cover the high pressure relic minerals (fig. 1a-c), the ± 2 kb large error bars of the P-T calculations resulting from low dwell time analyses (100 ms/pixel in imaging mode) prevent drawing definitive conclusions. However, the precision of chemical maps on major elements is good enough to highlight different generations of phyllosilicates. P-T maps of crystallization have been derived from these chemical maps where the location of HP relics could be inferred out of the background noise (fig. 1e). A new set of high precision (1 minute dwell time) point analyses performed on tiny grains pre-localized on P-T maps eventually revealed the HP history of rocks.

3. Necessity for high flux fluorescence microprobes

Using modern synchrotron XRF microprobe would have been of great interest in the WAC study framework, because the high flux delivered by beamlines provides analytical precision comparable to 1 minute long point EPMA analyses, but with dwell time in the millisecond range. In the case of latest generation synchrotron sources like NSLS-II, involving state of the art fluorescence detectors, mapping of elements with sup-ppm concentrations would be feasible with millisecond dwell-time, providing that detectors are able to handle high count rates. Trace elements are of great interest in Earth Science because they provide geologists with information on large geodynamic cycles. Many trace elements also are of economic interest. Being able to map their spatial distribution in rocks will bring new light on their transport modes. The SRX beamline at NSLS-II will be equipped with two kinds of energy dispersive X-ray spectrometers (EDX). One is the Maia[17-19], a collaborative undertaking between CSIRO and Brookhaven National Laboratory. The last version of the Maia benefits from a large solid angle of 1.2 sr with 384 elements that can handle dwell time in the 10's μsecond range. With a maximum count rate of 50k counts/s/element, Maia can manage a maximum rate of 19M counts/s before saturating. The drawback of the detector is currently its average energy resolution of about 300 eV. Figures 3 and 4 show images of trace elements concentrations in a metamorphic rock containing quartz, chloritoid and garnet in addition to chl and phg[20]. Maps were acquired with the 384 version B of Maia on the P06 beamline at Petra III, with incoming beam energy set at 20.5 keV. The experimental setup uses Kirkpatrick-Baez (KB) mirrors as focusing optics with characteristics close to those of SRX (fig. 2). During the experiment, the KB focusing optics provided about 10^{12} ph/s into a 300×300 nm^2 beam spot[21]. A 4×8 mm^2 area was scanned with a 0.6 μm large beam and a 0.5 ms dwell time. The images correspond to 13332×6666 $pixel^2$ hyperspectral fluorescence data from which elemental composition maps were extracted using the Dynamic Analysis method[22]. The method is based on a fundamental parameters approach to μXRF quantification using the GeoPIXE software that builds a transformation matrix to project individual events into pixel concentration after following the fitting of a representative spectrum with emission lines as well as the Compton and Thomson scattering[17]. The resolution (in term of pixel number) obtained here is 43 times higher than the Blu-ray standard for instance. For such analytical conditions, detection limits are of the order of a few ppm. But it must be kept in mind that the ability to discern spatial chemical gradients also depends on relative chemical variations. The quantitative analyses forming the images of figures 3 and 4 are being studied in the framework of a large project concerning trace elements in metamorphic rocks and are not discussed here. They are exhibited here to illustrate the results scientists can expect with sub-millisecond dwell time for different trace element contents. Imaging tests were also performed on an EPMA Cameca SX100 at the ISTEP laboratory of the Paris 6 university in France.

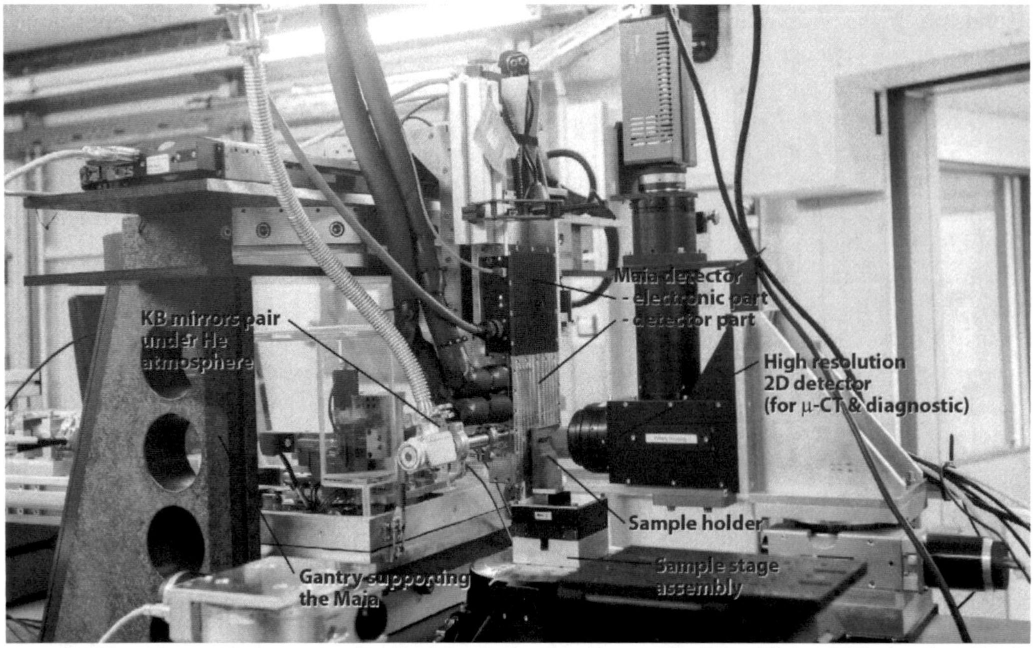

Figure 2. Experimental setup at the P06 beamline at Petra III. The Maia (384 version B) is in introduced between the KB focusing optics and the sample.

For analytical condition of 20 kV and 300 nA, a detection limit of 400-500 ppm has been reached with a dwell time of 300 ms using 4 spectrometers (~1.2 ms dwell time / spectrometer).

SRX will also operate with one ME3 and another ME4 multi-element Vortex silicon drift detector (50 mm^2 per element). Detectors will be combined with the new Quantum Xspress3 electronics, which claims a high count-rate capability providing 200, 300 and 400 eV energy resolution for photon flux of 1, 2.3 and 3 million counts per second (Mn Kα emission at ~5.9 keV). This means that the 7 elements should be able to provide 7, 16 or 21 million counts/s for energy resolution of 200, 300 or 400 eV, although a sub-ms pixel spectrum acquisition capability is yet to be demonstrated. However, the issue that remains when large elements experience high count rate is the pile-up effect for non-diluted samples like rocks. Commissioning data shows that with 3 M counts/s/element throughput on concentrated samples, background level generated with pile-up could reach about 20% of the main peaks intensity, jeopardizing trace elements detection. Z minus 1 filters have to be introduced between the sample and the detector to preferentially attenuate the most intense emission lines. For instance, a Mn foil with an absorption edge at 6.5 keV will absorb preferentially Fe K-α emission line. To summarize, Maia provides very high throughput, allowing raster scanning large areas of a square centimeter with submicron resolution and sensitivity of the order of a ppm. It can discern subtle chemico-spatial variations in a rock for a large range of elements and concentrations or locate tiny and rare particles[23] or preserved relic mineral in matrix. It also opens new paths for fluorescence μ-tomography or μ-XANES imaging spectroscopy. Megapixel hyperspectral XANES images so far only are achievable using full-field absorption imaging[24-26] can be obtained in fluorescence mode[27]. New detectors, such as the Vortex multi element arrays coupled with fast electronics also show great promise given a suitable data acquisition system.

Without considering potential radiation damages on the sample, would these detectors be able to fully sustain the outstanding flux of SRX? SRX will host in its end-station a microprobe and a nanoprobe, focusing on the sample respectively more than 10^{13} ph/s into a submicron spot and more than 10^{12} ph/s into a 50 nm spot. Both should be the most intense submicron and sub-100 nanometer

Journal of Physics: Conference Series **499** (2014) 012012 doi:10.1088/1742-6596/499/1/012012

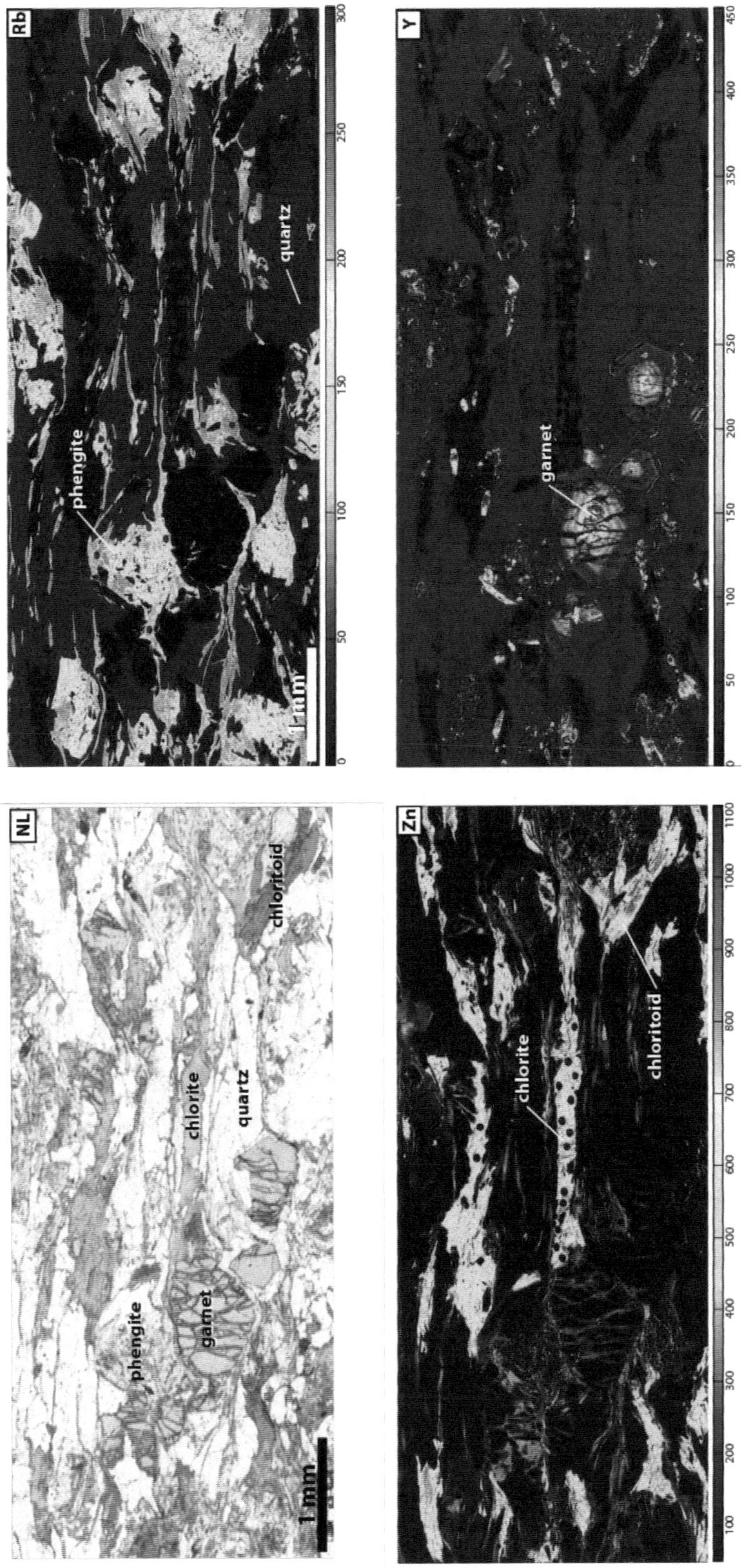

Figure 3. Quantitative images of Rb, Zn and Y concentrations acquired together with the Cr image showed in fig.4.

Journal of Physics: Conference Series **499** (2014) 012012 doi:10.1088/1742-6596/499/1/012012

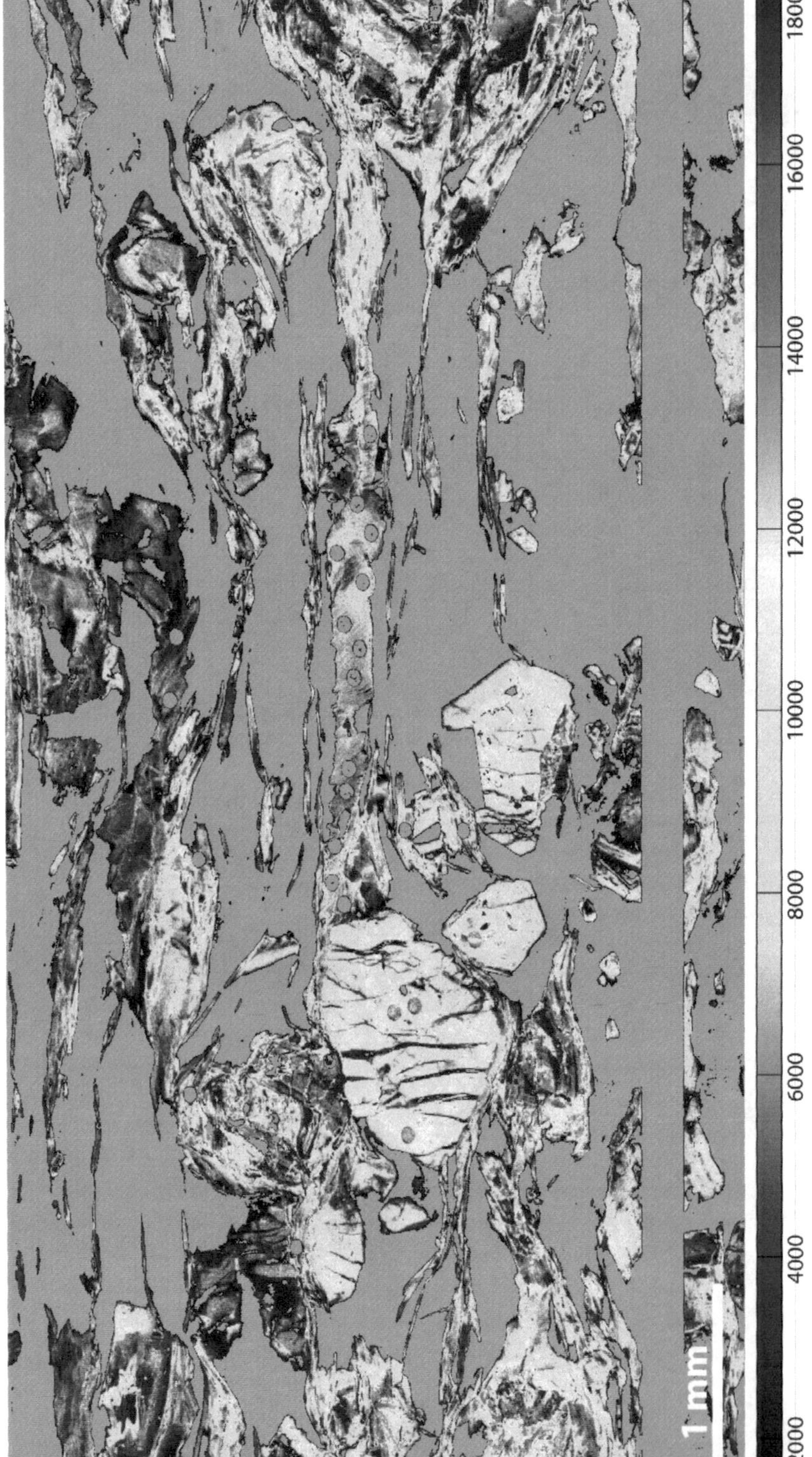

Figure 4. Quantitative image of Cr concentration (ppm) in a qtz-chl-phg-cld-gt-accessory minerals bearing metamorphic rock. This image has been acquired with the Maia 384 elements detector on the P06 beamline at Petra III. Some of the minerals present in the rock are labelled in fig.3. The disk shape marks observable in minerals are the consequence of former ICP-MS analyses

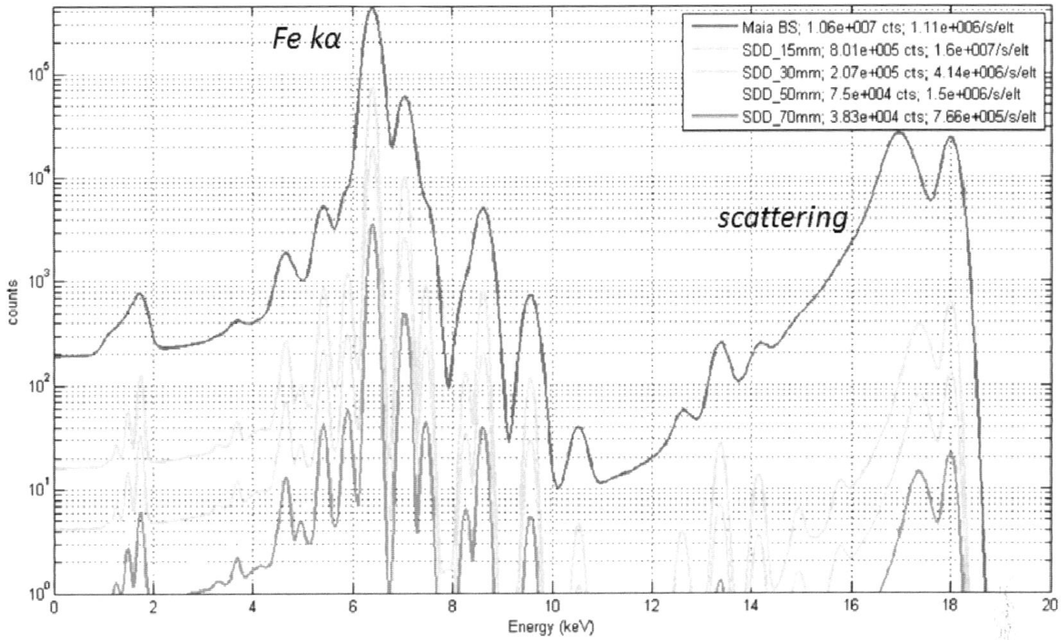

Figure 5. Fluorescence spectra simulations derived from the XMI-MSIM software[25], for a Maia detector in backscattered geometry and a single element SDD for different working distance (15, 30, 50 and 70 mm), when operating on the SRX nanoprobe, i.e. with an incoming flux of 10^{12} ph/s. Dwell time: 50 ms. The Maia saturates at 5.10^4 ph/s, and the SDD warranty an energy resolution of 200, 300 and 400 eV for count rates of 1, 2.3 and 3M ph/s.

beams of the world. When analyzing non diluted samples, with intense major element lines, the experiment performed at Petra III on P06 beamline demonstrated that the incoming beam intensity should stay below 10^{11} ph/s to prevent detector saturation. To anticipate performances of the SRX detectors, simulations of fluorescence spectra were performed with the XMI-MSIM software[28], a Monte Carlo code dedicated to ED-XRF spectrometers including several features such as the simulation of M-lines and cascade effects. Simulations represent the Maia 384 and a single SDD element (energy resolution fixed at 200 eV) for different working distances and an incoming beam of 10^{12} ph/s as for the nanoprobe (fig. 5). The simulated sample is a 20 μm thick chl crystal. The input composition corresponds to an analyses made by laser ablation mass spectroscopy on a chl imaged in fig. 3-4. It turns out that the Maia will experience 10^6 ph/s/elt while it can only handle 5.10^4 ph/s/elt, and the SDD should move 6 cm back to operate at 1Mcts/s. Current state of the art EDX technologies are clearly a limiting factor for operating SRX at its full potential. This high flux density however offers flexibility. It allows to maintain short dwell time even for highly diluted samples, or still to work on very thin samples (<1 μm). In such cases, the amount of fluorescence emitted by thin samples drops while the spatial resolution degradation due to X-ray depth penetration is minimized.

Utilization of new wavelength dispersive spectrometers (WDS) is a credible alternative to EDX in order to benefit from the very intense beams of ultra-low emittance new synchrotron sources. With recent progresses in fabrication of polycapillary optics, compact WDS with large solid angles are now feasible and easily integrated into fluorescence microprobes[29, 30]. They provide energy resolution superior to EDX, ranging from ~5 to ~40 eV according to photon energy probed and crystal analyzers used for Bragg reflections. If these polycapillary optics based WDS designs still not sustain the comparison with recent EDX in term of efficiency, it would not saturate and not produce any pile-up with the 10^{13} ph/s flux of the SRX microprobe, and would then outclass the EDX presented earlier. However, the down side of WDS is that a full emission spectrum cannot be acquired in one shot. A continuous spectrum is obtained by rocking the analyzer and the detector. It means that one set polycapillary / analyzer / detector at fixed location in imaging mode can analyze only one element at a time.

4. Zooming from micro to nanometer scale with XRF analyses

Fluorescence nanoprobes are nowadays blooming in synchrotron facilities, especially to support research on nanomaterials, but these instruments are also relevant for metamorphic geology where tiny objects or nanoscale processes are involved. For instance, the study of small mineral inclusions, whose crystallization occurs at high P-T conditions, presents a strong interest since such objects are a window on Earth interior. Inclusions trapped in diamonds can be exhumed from locations as deep as the lower mantle[31]. Because of technical limitations, most studies have been performed so far on micro inclusions, but nanoprobes make possible to revisit samples containing nano-inclusions[32].

Nanometric spatial resolution is also particularly relevant for probing diagenetic to low grade metamorphic rocks. Indeed, mineralogical processes generally occur at the sub-micron level for relatively low temperature ($T<250^{o}C$), when slow kinetics of reactions do not permit crystallization of large grains. Phg crystallize as clay-size illite, intercalations of clays with oxides are observed in micrometric volumes. In the vein of the study described in paragraph 2, Bourdelle *et al.* interested in the compositional variations of sandstones, but using high resolution analyses acquired with transmission electron microscope (TEM) coupled with EDX[4]. As for phyllosilicate of West Africa greenstones, flakes of illite and chl contained in (low T diagenetic conditions) sandstones from the Gulf Coast in Texas exhibit compositional variations but for spatial range spanning a couple of hundreds of nanometres only. Crystals present rims around relic cores overgrowing during diagenetic processes. From these fine scale analyses, it is then possible to apply newly developed thermobarometers involving clays[4, 11, 12] and low T chl[33, 34] to relate measured compositions to P-T conditions.

X-ray nanoprobes are still not able to compete with the sub-nanometric resolution of TEM instruments. However, obtaining reliable quantitative chemical analyses of phyllosilicate with TEM-EDX implies to reduce the power density of the beam by defocusing electrons from1 to 50 nm[4]. In this case, the expected 30 nm large X-ray beam of the SRX nanoprobe surpasses the resolution provided by TEM. More important, the acquisition time needed to get a decent signal with TEM-EDX is in the minute range. It completely annihilates imaging possibilities while on the SRX nanoprobe, dwell times can be maintained in the millisecond range to acquire high resolution large images of trace elements concentration. Also, the SRX nanoprobe will have the unique opportunity to perform XANES spectroscopy at the nanoscale with achromatic and high efficiency KB optics, offering the possibility to assess oxidation state of elements or to identify mineralogical phases with the XANES spectra "finger print" approach.

To conclude, coupling of nanoscale analyses with suitable thermobarometers offers new perspectives. In the case of metamorphic rocks, this approach reveals the end of their retrograde P-T paths (<250 ^{o}C), i.e. the final steps of exhumation. For industrial applications, a more accurate modeling of sedimentary basins with these mineralogical tools can substantially enhance prospection efficiency of petroleum, uranium or other precious ore deposits.

5. Radiation damage considerations

Radiation damage on organic materials has been intensively studied since such compounds are easily altered by intense X-ray beams. However, effects of X-rays on hard condense matter can be observed as well, especially when submicron beams with high power density are used. Regarding SRX, the microprobe and the nanoprobe will subject samples to very high power density of about 100 and 1000 kW/mm^2 (corresponding to 1.92 mW in the nanobeam) at 12 keV. The use of such high fluence X-ray beams will necessitate extreme caution when analyzing materials to avoid any misinterpretation derived from analytical artifacts. One can cite as an example the case of S^{4+} species observed in silicate glasses. S^{4+} were first documented in natural melt inclusions in olivine of arc-related magmas[35, 36] before being identified as an instrumental artifact caused by irradiation with intense focused X-ray beams[37].

The main interaction process between X-ray beams and matter is photo-ionization yielding photoelectrons with energies lower or equal to the incident beam energy, followed by inelastic (Compton) and elastic (Rayleigh) scattering. The predominant relaxation process following the ionization is through Auger electron emission. In silicates, the travel range of electrons generated by X-ray beam below 20 keV is of the order of 1 to 2 μm. Therefore, except from a small fraction re-emitted by fluorescence or scattered, most of the absorbed energy remains in the sample[38]. Radiation damage results in oxidation state change of elements as well as mass or crystallinity loss. These effects required mechanisms for converting energy acquired by atomic electrons to kinetic energy[39]. In detail, these complex mechanisms are not always well understood and vary according to the targeted materials, the thickness of the sample, the environmental conditions, and the power density of the beam. For instance, Alkali halides are among the most radiation sensitive inorganic material whereas oxides or transition metal oxides can sustain a much higher dose before being altered by electrostatic charging[39]. In glass silicate, the use of a sub-microbeam focused with a zone plate induces photo-reduction of sulfur with the decrease S^{6+} in favor to S^{4+}. When submitted to flux two orders of magnitude higher (use of KB as focusing optics), the same sample exhibits this time photo-oxidation. Fine structure variations in a series of XANES spectra accumulated at the same sample location indicates that the S oxidation state change is coupled with reorganization of the S local structural environment[37].

Simionovici *et al.*[38] discussed radiation damage observed in silicates of the *Stardust* return samples mission. 2 μm large Mg- and Al-rich silicate grains trapped in aerogel were analyzed at the ESRF on ID22NI with a 150 nm large pink beam with a flux of 3×10^{10} ph/s. The dose deposited with 200 ms dwell time per pixel did not damage the samples, but a comparison of subsequent XRF images of the same area acquired 8 s dwell time showed smearing of the grains along the fast scan direction and mass loss involving Mn and K. The identified damage mechanism is electrostatic charging subsequent to photoelectron and Auger electron emissions, while ionic displacement and heating processes are negligible for the analytical conditions. The power density of the SRX nanoprobe is however about 350 times higher than the one of ID22NI. It means that the dose responsible of the silicates grains radiolysis would be reached at SRX with 25 ms dwell time. In addition, to validate the assumption that heating has still negligible effects on such materials with 1 MW/mm^2, FEA are currently ongoing to establish how much the temperature would increase at the sample, either in air or in vacuum with cryogenic conditions.

After these considerations, it is clear that SRX needs a strategy to manage radiation damage even for inorganic materials. First of all, particular care must be brought to the most fragile materials, like the ones composed of alkaline elements. Hydrated materials are also very sensitive. Photoelectrons hydrolyze water making free radicals that will as a second effect produce oxidative attack. Clays and phyllosilicates comprising both alkali and water in the interlayer would be particularly subject to radiation damage. Some procedure or special sample preparation can help to circumvent this problem. For instance, in paragraph 3 is discussed the interest of working on FIB sections to minimize the spatial resolution loss due to X-ray depth penetration while decreasing the throughput on the detectors. Using sub-micron thick thin sections would also minimize the amount of absorbed Auger electrons and therefore of radiation damage. Also, when working in air, intense X-ray beams break some O_2 molecules bonds. The free O atoms recombined then with O_2 to form ozone[40]. Placing the sample in N or He atmosphere prevents the production of ozone that is a strong oxidant able to corrode sample surfaces. One general approach that SRX could adopt to deal with radiation damage could be to shorten dwell times to sub-ms levels, taking benefit of the very fast fluorescence detectors of SRX. Several frames of the same area could be acquired with 0.2 ms dwell time. This may allow enough time for charge migration and a lessening of the charge trapped gradients that drive some element loss effects (e.g. alkalis). XANES spectroscopy is a very valuable diagnostic tool for radiation damage assessment. With the help of filters, low flux XANES spectra can be measured before and after image or stationary acquisitions to track potential oxidation state (e.g. transition metal in oxides) or crystallinity changes identifiable by fine spectral modifications. In the same way, fast images with moderate flux could be performed in a first time. Comparison of peak ratio between images acquired with low and high power density would help identifying potential mass loss.

Because the degree of radiation damage experienced by materials is composition dependent, an early study of standard samples would benefit to future experiments performed at SRX. In a certain extent, those results could also be extended to Free Electron Laser instruments since interactions between hard condense matter and high fluence X-ray beam are still poorly known.

Conclusion

The use of elemental chemical imaging derived from XRF microprobes were an important milestone for the understanding of mineralogical and geodynamic processes, as is illustrated in the study of greenstones from WAC. The steady improvement of these instruments realized by the current construction or upgrade of ultra-bright diffraction limited synchrotron sources like NSLS-II opens new horizons for high speed chemical imaging of trace elements with large fields of view. Examples of acquisitions performed with the Maia 384 on a metamorphic rock underscore the achievable performances with current beamlines and EDX detectors. Simulations also show that further detector development is necessary to fully benefit from the potential of a beamline like SRX. In addition, the recent improvement in optics fabrication enables construction of efficient nanoprobes that are of interest for applications concerning low T metamorphism. However, the use of the high fluence beam of SRX necessitates strategy to control and diagnose radiation damage even on hard condense matter.

Acknowledgements

This work was financially supported by the NSLS-II project, itself funded by the US department of Energy, Office of Science and Office of Basic Energy Sciences. We are grateful to Ulrike Boesenberg and Gerd Wellenreuther from the P06 beamline team at Petra III, who helped setting up the experiment aiming to test the new Maia 384 detector. We also want to thank Robin Kirkham & Pete Siddons for their support and advices concerning the Maia. At last, thanks to Tom Schoonjans for his support in configuring the XMI-MSIM software and the SRX team endeavoring to build an outstanding instrument.

References

[1] P. Yamato, P. Agard, B. Goffé, V. De Andrade, O. Vidal, L. Jolivet, New, high-precision P-T estimates for Oman blueschists: implications for obduction, nappe stacking and exhumation processes, Journal of Metamorphic Geology, 25 (2007) 657.
[2] O. Vidal, V. De Andrade, E. Lewin, M. Muñoz, T. Parra, S. Pascarelli, P-T-deformation-Fe3+/Fe2+ mapping at the thin section scale and comparison with XANES mapping: application to a garnet-bearing metapelite from the Sambagawa metamorphic belt (Japan), Journal of Metamorphic Geology, 24 (2006) 669.
[3] V. De Andrade, O. Vidal, E. Lewin, P. O'Brien, P. Agard, Quantification of electron microprobe compositional maps of rock thin sections: an optimized method and examples, Journal of Metamorphic Geology, 24 (2006) 655.
[4] P.T. Bourdelle F., Beyssac O., Chopin C., Vidal O., Clay minerals as geo-thermometer: A comparative study based on high spatial analyses of illite and chlorite in Gulf Coast sandstones (Texas, U.S.A.), American Mineralogist, 98 (2013) 914-926.
[5] O. Vidal, T. Parra, Exhumation paths of high-pressure metapelites obtained from local equilibria for chlorite–phengite assemblages, Geological Journal, 35 (2000) 139-161.
[6] P. Lanari, O. Vidal, V. De Andrade, B. Dubacq, E. Lewin, E.G. Grosch, S. Schwartz, XMapTools: A MATLAB©-based program for electron microprobe X-ray image processing and geothermobarometry, Computers & Geosciences, (2013).
[7] E.S. Reich, Ultimate upgrade for US synchrotron, Nature, 501 (2013) 148-149.

[8] V. De Andrade, J. Thieme, P. Northrup, Y. Yao, A. Lanzirotti, P. Eng, Q. Shen, The sub-micron resolution X-ray spectroscopy beamline at NSLS-II, Nuclear Instruments and Methods in Physics Research Section A: Accelerators, Spectrometers, Detectors and Associated Equipment, 649 (2011) 46-48.
[9] V. De Andrade, J. Thieme, O. Chubar, M. Idir, Simulation and optimization of the NSLS-II SRX beamline combining ray-tracing and wavefront propagation, SPIE, In press (2011).
[10] J. Ganne, V. De Andrade, R.F. Weinberg, O. Vidal, B. Dubacq, N. Kagambega, S. Naba, L. Baratoux, M. Jessell, J. Allibon, Modern-style plate subduction preserved in the Palaeoproterozoic West African craton, Nature Geosci, 5 (2012) 60-65.
[11] B. Dubacq, O. Vidal, V. De Andrade, Dehydration of dioctahedral aluminous phyllosilicates: thermodynamic modelling and implications for thermobarometric estimates, Contributions to Mineralogy and Petrology, 159 (2010) 159-174.
[12] O. Vidal, B. Dubacq, Thermodynamic modelling of clay dehydration, stability and compositional evolution with temperature, pressure and H2O activity, Geochimica et Cosmochimica Acta, 73 (2009) 6544-6564.
[13] T. Parra, O. Vidal, P. Agard, A thermodynamic model for Fe–Mg dioctahedral K white micas using data from phase-equilibrium experiments and natural pelitic assemblages, Contrib Mineral Petrol, 143 (2002) 706-732.
[14] O. Vidal, T. Parra, F. Trotet, A Thermodynamic Model for Fe-Mg Aluminous Chlorite Using Data from Phase Equilibrium Experiments and Natural Pelitic Assemblages in the 100° to 600°c, 1 to 25 kb Range, American Journal of Science, 301 (2001) 557-592.
[15] R.G. Berman, Thermobarometry using multi-equilibrium calculations: a new technique, with petrological applications, Canadian Mineralogist, 29 (1991) 833-855.
[16] M. Muñoz, V. De Andrade, O. Vidal, E. Lewin, S. Pascarelli, J. Susini, Redox and speciation micromapping using dispersive X-ray absorption spectroscopy: Application to iron in chlorite mineral of a metamorphic rock thin section, Geochem. Geophys. Geosyst., 7 (2006) Q11020 DOI 11010.11029/12006GC001381.
[17] C.G. Ryan, D.P. Siddons, R. Kirkham, P.A. Dunn, A. Kuczewski, G. Moorhead, G. De Geronimo, D.J. Paterson, M.D. de Jonge, R.M. Hough, M.J. Lintern, D.L. Howard, P. Kappen, J. Cleverley, The New Maia Detector System: Methods For High Definition Trace Element Imaging Of Natural Material, AIP Conference Proceedings, 1221 (2010) 9-17.
[18] C.G. Ryan, D.P. Siddons, G. Moorhead, R. Kirkham, P.A. Dunn, A. Dragone, G. De Geronimo, Large detector array and real-time processing and elemental image projection of X-ray and proton microprobe fluorescence data, Nuclear Instruments and Methods in Physics Research Section B: Beam Interactions with Materials and Atoms, 260 (2007) 1-7.
[19] C. Ryan, R. Kirkham, R. Hough, G. Moorhead, D. Siddons, M. De Jonge, D. Paterson, G. De Geronimo, D. Howard, J. Cleverley, Elemental X-ray imaging using the Maia detector array: The benefits and challenges of large solid-angle, Nuclear Instruments and Methods in Physics Research Section A: Accelerators, Spectrometers, Detectors and Associated Equipment, 619 (2010) 37-43.
[20] A. Plunder, P. Agard, B. Dubacq, C. Chopin, M. Bellanger, How continuous and precise is the record of P–T paths? Insights from combined thermobarometry and thermodynamic modelling into subduction dynamics (Schistes Lustrés, W. Alps), Journal of Metamorphic Geology, 30 (2012) 323-346.
[21] C.G. Schroer, P. Boye, J.M. Feldkamp, J. Patommel, D. Samberg, A. Schropp, A. Schwab, S. Stephan, G. Falkenberg, G. Wellenreuther, Hard X-ray nanoprobe at beamline P06 at PETRA III, Nuclear Instruments and Methods in Physics Research Section A: Accelerators, Spectrometers, Detectors and Associated Equipment, 616 (2010) 93-97.
[22] C. Ryan, Quantitative trace element imaging using PIXE and the nuclear microprobe, International Journal of Imaging Systems and Technology, 11 (2000) 219-230.
[23] M. Lintern, R. Anand, C. Ryan, D. Paterson, Natural gold particles in Eucalyptus leaves and their relevance to exploration for buried gold deposits, Nat Commun, 4 (2013).

[24] B. Fayard, E. Pouyet, G. Berruyer, D. Bugnazet, C. Cornu, C. Cotte, V. De Andrade, F. Di Chiaro, O. Hignette, J. Kieffer, T. Martin, E. Papillon, M. Salomé, V.A. Sole, The new ID21 XANES full-field end-station at ESRF, Journal of Physics: Conference Series, 425 (2012) 192001.
[25] P. Beck, V. De Andrade, F.R. Orthous-Daunay, G. Veronesi, M. Cotte, E. Quirico, B. Schmitt, The redox state of iron in the matrix of CI, CM and metamorphosed CM chondrites by XANES spectroscopy, Geochimica et Cosmochimica Acta, (2012).
[26] V. De Andrade, J. Susini, M. Salomé, O. Beraldin, C. Rigault, T. Heymes, E. Lewin, O. Vidal, Submicrometer Hyperspectral X-ray Imaging of Heterogeneous Rocks and Geomaterials: Applications at the Fe K-Edge, Analytical Chemistry, 83 (2011) 4220-4227.
[27] N.J. Cook, C.L. Ciobanu, J. Brugger, B. Etschmann, D.L. Howard, M.D. de Jonge, C. Ryan, D. Paterson, Determination of the oxidation state of Cu in substituted Cu-In-Fe-bearing sphalerite via μ-XANES spectroscopy, American Mineralogist, 97 (2012) 476-479.
[28] T. Schoonjans, V.A. Solé, L. Vincze, M. Sanchez del Rio, K. Appel, C. Ferrero, A general Monte Carlo simulation of energy-dispersive X-ray fluorescence spectrometers â€" Part 6. Quantification through iterative simulations, Spectrochimica Acta Part B: Atomic Spectroscopy, 82 (2013) 36-41.
[29] M. Cotte, J. Szlachetko, S. Lahlil, M. Salome, V.A. Sole, I. Biron, J. Susini, Coupling a wavelength dispersive spectrometer with a synchrotron-based X-ray microscope: A winning combination for micro-X-ray fluorescence and micro-XANES analyses of complex artistic materials, Journal of Analytical Atomic Spectrometry, 26 1051-1059.
[30] J. Szlachetko, M. Cotte, J. Morse, M. Salome, P. Jagodzinski, J.-C. Dousse, J. Hoszowska, Y. Kayser, J. Susini, Wavelength-dispersive spectrometer for X-ray microfluorescence analysis at the X-ray microscopy beamline ID21 (ESRF), Journal of Synchrotron Radiation, 17 (2010) 400-408.
[31] M.J. Walter, S.C. Kohn, D. Araujo, G.P. Bulanova, C.B. Smith, E. Gaillou, J. Wang, A. Steele, S.B. Shirey, Deep Mantle Cycling of Oceanic Crust: Evidence from Diamonds and Their Mineral Inclusions, Science, 334 (2011) 54-57.
[32] A.M. Logvinova, R. Wirth, E.N. Fedorova, N.V. Sobolev, Nanometre-sized mineral and fluid inclusions in cloudy Siberian diamonds: new insights on diamond formation, European Journal of Mineralogy, 20 (2008) 317-331.
[33] F. Bourdelle, T. Parra, C. Chopin, O. Beyssac, A new chlorite geothermometer for diagenetic to low-grade metamorphic conditions, Contrib Mineral Petrol, 165 (2013) 723-735.
[34] A. Inoue, A. Meunier, P. Patrier-Mas, C. Rigault, D. Beaufort, P. Vieillard, Application of chemical geothermometry to low-temperature trioctahedral chlorites, Clays and Clay Minerals, 57 (2009) 371-382.
[35] N. Métrich, M. Bonnin-Mosbah, J. Susini, B. Menez, L. Galoisy, Presence of sulfite (SIV) in arc magmas: Implications for volcanic sulfur emissions, Geophysical Research Letters, 29 (2002) 33-31-33-34.
[36] M. Bonnin-Mosbah, N. Métrich, J. Susini, M. Salome, D. Massare, B. Menez, Micro X-ray absorption near edge structure at the sulfur and iron K-edges in natural silicate glasses, Spectrochimica Acta Part B: Atomic Spectroscopy, 57 (2002) 711-725.
[37] M. Wilke, P.J. Jugo, K. Klimm, J. Susini, R. Botcharnikov, S.C. Kohn, M. Janousch, The origin of S4+ detected in silicate glasses by XANES, American Mineralogist, 93 (2008) 235-240.
[38] A. Simionovici, C. Allen, S. Bajt, R. Bastien, H. Bechtel, J. Borg, F. Brenker, J. Bridges, D. Brownlee, M. Burchell, High Fluence Synchrotron Radiation Microprobe Effects on Stardust Interstellar Dust Candidates, Lunar and Planetary Institute Science Conference Abstracts, 2011, pp. 2812.
[39] R.F. Egerton, P. Li, M. Malac, Radiation damage in the TEM and SEM, Micron, 35 (2004) 399-409.
[40] C. Weilandics, N. Rohrig, N.F. Gmür, Ozone production due to synchrotron radiation, Nuclear Instruments and Methods in Physics Research Section A: Accelerators, Spectrometers, Detectors and Associated Equipment, 266 (1988) 691-698.

22nd International Congress on X-Ray Optics and Microanalysis IOP Publishing
Journal of Physics: Conference Series **499** (2014) 012013 doi:10.1088/1742-6596/499/1/012013

Non-negative factor analysis supporting the interpretation of elemental distribution images acquired by XRF

Matthias Alfeld[1], Mirwaes Wahabzada[2], Christian Bauckhage[2], Kristian Kersting[2,3], Gerd Wellenreuther[4], Gerald Falkenberg[1]

[1] Photon Science, DESY, Notkestraße 85, 22607 Hamburg, Germany
[2] Fraunhofer IAIS, Schloss Birlinghoven, 53757 Sankt Augustin, Germany
[3] Computer Science Department, TU Dortmund University, Joseph-von-Fraunhofer-Str. 23, 44227 Dortmund, Germany
[4] European XFEL GmbH, Albert-Einstein-Ring 19, 22761 Hamburg, Germany

E-mail: `matthias.alfeld@desy.de`

Abstract.
Stacks of elemental distribution images acquired by XRF can be difficult to interpret, if they contain high degrees of redundancy and components differing in their quantitative but not qualitative elemental composition. Factor analysis, mainly in the form of Principal Component Analysis (PCA), has been used to reduce the level of redundancy and highlight correlations. PCA, however, does not yield physically meaningful representations as they often contain negative values. This limitation can be overcome, by employing factor analysis that is restricted to non-negativity. In this paper we present the first application of the Python Matrix Factorization Module (pymf) on XRF data. This is done in a case study on the painting *Saul and David* from the studio of Rembrandt van Rijn. We show how the discrimination between two different Co containing compounds with minimum user intervention and *a priori* knowledge is supported by Non-Negative Matrix Factorization (NMF).

1. Introduction

XRF imaging is one of the most versatile techniques for the acquisition of elemental distribution images. It is used with different instruments for the investigation of a wide range of objects. These objects range from microscopic samples, which are investigated with a lateral resolution of less than a micrometer, to historical paintings, of which several square meters can be imaged with a resolution of a fraction of a millimeter.

If in these samples components are present that contain multiple elements detectable by XRF, a high level of redundancy can be observed, as the lateral distribution of one component would be represented in multiple elemental distribution images. Further, components differing in their quantitative, but not qualitative elemental composition can be difficult to discriminate. For many years, factor analysis has been used to support the interpretation of XRF data by reducing redundancies and the dimensionality of a data matrix, composed of raw spectral data or elemental distribution images.

The underlying problem is shown in Figure 1. Factor analysis aims at the decomposition of a matrix $\mathbf{V} \in \mathbb{R}^{d \times n}$ into two lower ranked matrices $\mathbf{W} \in \mathbb{R}^{d \times k}$ and $\mathbf{H} \in \mathbb{R}^{k \times n}$. Here, the

Journal of Physics: Conference Series **499** (2014) 012013 doi:10.1088/1742-6596/499/1/012013

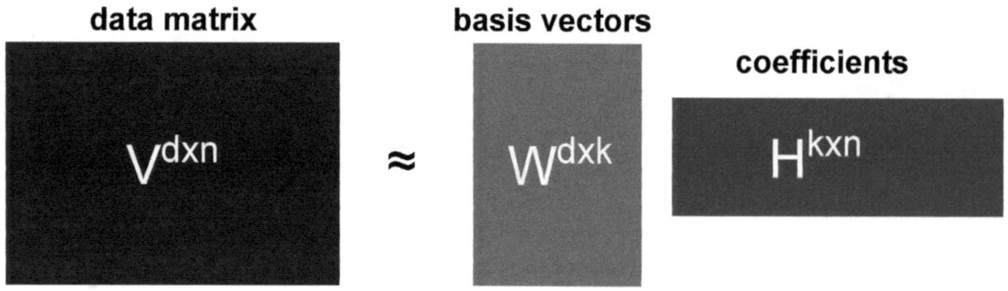

Figure 1. Factor Analysis: The data matrix **V** is described as the product of the basis vectors **W** and the coefficients **H**.

matrix **V** represents an XRF data matrix, whose columns (**n**) correspond to pixels represented by the recorded intensity of **d** sets of fluorescence lines of different elements. The basis vectors **W** describe the loading of the data set's variables on $\mathbf{k} < \mathbf{d}$ bases, while the coefficients in **H** represent the lateral distribution of the degree that bases describe the data matrix. They represent the coordinates of the input objects in a lower-dimensional space, spanned by the basis vectors in **W**. A common approach for the factorization is Principal Component Analysis (PCA) that has been used for factor analysis in the interpretation of XRF data sets for a long time [1]. PCA is very effective in terms of data compression and noise reduction. Here, an approach is to compute the eigenvectors of the data covariance matrix, with largest eigenvalues as basis vectors **W**. Here **W** is called the loading matrix and **H** the score matrix.

But although very effective, it has been criticized for the lack of interpretability of the resulting basis vectors. XRF data consists only of non-negative values and it is desirable that the resulting representation retains this property, which is not the case for PCA. To overcome this limitation other approaches constrain the resulting factorization to non-negativity. Multivariate statistical analysis with non-negative constrains is well known for energy dispersive X-ray spectroscopy with electron beams [2, 3]. In this paper we employ Non-Negative Matrix Factorization (NMF) [4]. It is worth noting, if $\mathbf{k} = \mathbf{d}$, PCA yields an exact representation of the data matrix **V** (we only rotate the data along principal components), while NMF only provides an approximation. However, while a good representation of the data set is desirable, the real criterion for the quality of the decomposition of the data set is how well the different components are represented and separated.

NMF is a widely used factor analysis method with non-negative constrains that requires also as input a matrix with non-negative values. It defines simple update rules for the matrices **W** and **H** with non-negative constraints, based on the squared Frobenius norm or Kullback-Leibler Divergence as objective functions (see [4] for more details). NMF results in a sparse and part based representation, which is more interpretable than a PCA decomposition. A comparison of basis vectors are shown in Figure 2, where we applied these methods on the CBCL face database[1].

Each image represents a basis vector (column in matrix **W**). Using NMF on this test data set aims at the decomposition of the faces into their different basic components (e.g. nose, ear, eyes). This follows from the constraints that allow only additive combinations and thus results in part based representations, as shown in Figure 2 (right). Therefore, they allow an easier interpretation of bases. In contrast, basis images found by PCA (left) can also contain negative entries (red parts) that are not easy to understand.

Other variations of NMF increase the interpretability using convexity constraints [5, 6].

[1] http://cbcl.mit.edu/software-datasets/FaceData.html

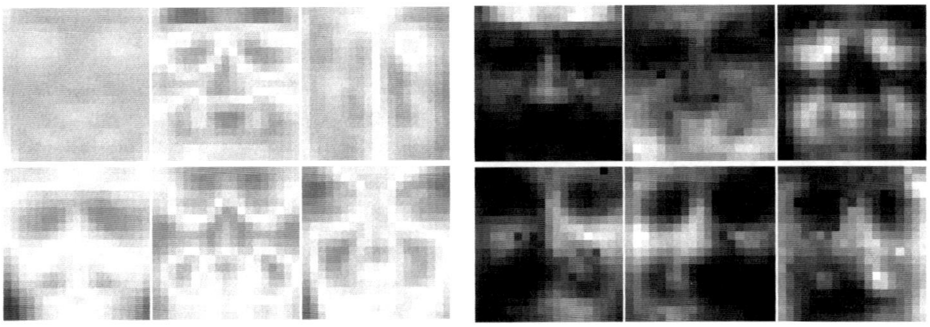

Figure 2. Resulting basis vectors (columns in **W**) for PCA (left) and NMF (right) from the CBCL face database. The data set is represented by a matrix, whose columns correspond to one particular face image. PCA basis vectors (eigenfaces) contain negative values (red parts) whereas those found by NMF do not.

Alternatively, one can compute a factorization by representing the data as combination of few actual data points (see [7] and references therein). This further increase the interpretability of resulting basis vectors (columns in matrix **W**), since they represent real data points from matrix **V**, and compared to NMF they are not part based.

In this paper we apply NMF to a XRF data set acquired on the painting *Saul and David*, which is shown in Figure 3. The painting is part of the collection of the Mauritshuis Museum in The Hague (NL) and is attributed to the studio of Rembrandt van Rijn (1606-1669) with a disputed attribution to the master himself. The investigation of the central part of the painting aimed on the distinction between original materials and materials used in later conservation treatments. The investigation of the painting has been described in detail in [8].

2. The painting

At the moment of the investigation (early spring 2011), the painting was covered by a thick layer of darkened varnish and possible overpaint and was being studied prior to eventual conservation treatment. The planning for this was not straightforward, as the extent of original paint still present in the painting was unknown, due to its complex history. Moreover, the authenticity of the curtain depicted in the area between the figures was questioned.

At an unknown moment in the past the painting was cut into two separate pieces. The reason for this is unknown. Based on dimensions listed in the auction records the two figures were rejoined between 1830 and 1869. The join between the two pieces of canvas is clearly discernible in the X-ray radiograph (XRR , see Figure 3). Examination with the stereomicroscope revealed, that much of the curtain visible in the background of the painting was added after the parts were rejoined to visually blend them together. Further, it was known from investigations using a light microscope and SEM-EDX studies of paint cross sections that the original paint is still present under the surface of the curtain. However, its exact extent was unknown as these samples only provided local information. The original paint was found to be rich in smalt, a blue pigment used in the 17th century, consisting of ground Co-rich blue glass.

Later added paint layers can be identified by the presence of anachronistic compounds, which were not used in the 17th century. This could not be done by conventional techniques such as XRR and Infrared reflectography as the lead white ($(PbCO_3)_2Pb(OH)_2$) rich ground that also contains black pigments and the dark, bone black and earth pigment (Fe and Mn oxides) rich pictorial layers yield only low contrast images with these methods [8, 9].

Journal of Physics: Conference Series **499** (2014) 012013 doi:10.1088/1742-6596/499/1/012013

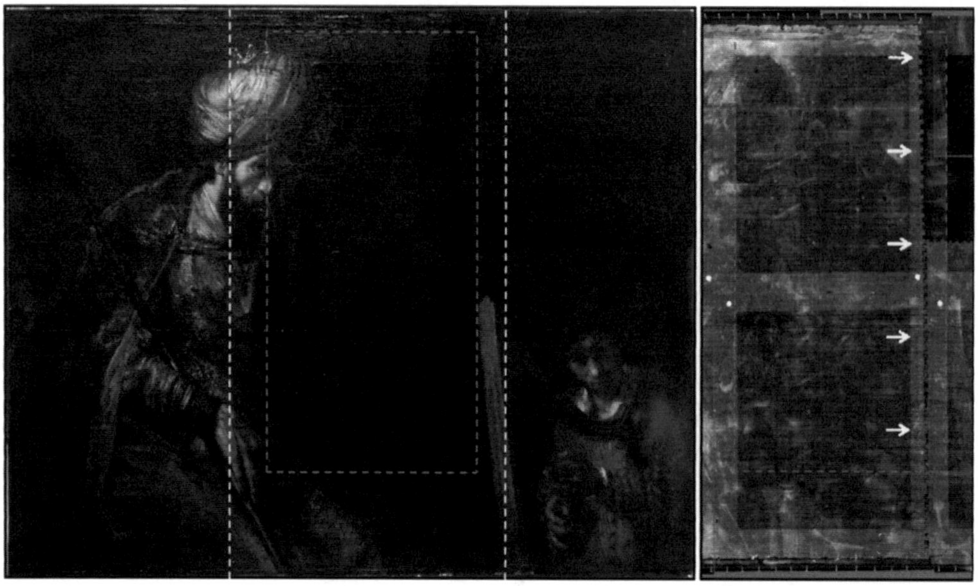

Figure 3. Rembrandt and/or Studio, Saul and David, c.1655, Mauritshuis inv. no. 621 / Bredius no. 526, oil on linen, 130.5 x 164.5 cm. In the XRR (right) the joins between the different pieces of canvas are clearly visible. The red dashed lines indicate the area investigated by MA-XRF, the yellow dashed ones the area of the painting shown in the XRR. The join is indicated in the XRR by yellow arrows.

3. Experimental

The painting was investigated in the conservation studio of the Mauritshuis Museum by means of a mobile XRF scanner. The scanner consists of a Mo-anode microfocus X-ray tube with a fixed polycapillary lens mounted with four energy dispersive SD-detectors on a motorized stage. This scanner allows for the investigation of an area of 60x25 cm^2 in a single scan. It is described in detail in [10] as "Instrument C". The painting was scanned with a step size of 1 mm and an average dwell time of 1.2 s per pixel, so that the central part of the painting (approx. 60x120 cm^2), which is discussed here, was investigated in roughly two weeks. The spectra of the four detectors were summed for each pixel prior to processing with the AXIL software package [11] and in-house written software.

4. Results

4.1. Elemental distribution images

In Figure 4 the elemental distribution images obtained from the central part of the painting are shown. Ca is present as a minor component in the black pigment bone black and as major constituent in chalk, which was used as a paint extender, and to fill the gaps in the join, probably in the form of chalk or gypsum. The distribution of Fe visualizes the earth pigments, which in general have a brownish/reddish color and were used in the original as well as in the later added paint. Since the curtain was mainly executed in these pigments, its folds are clearly visible in the Fe image. Mn is also present in earth pigments of darker tone. Cu is present in a green or blue pigment, which is present in the turban and to a minor amount in the ground of the painting. Pb is present in lead white. Hg is present in the red pigment vermilion (HgS) and was used in Saul's turban and to cover the join. Ti, Cr, Zn and Ba are mainly present in anachronistic materials, used to cover the joins. Cd is present in low abundance and it is unclear whether it

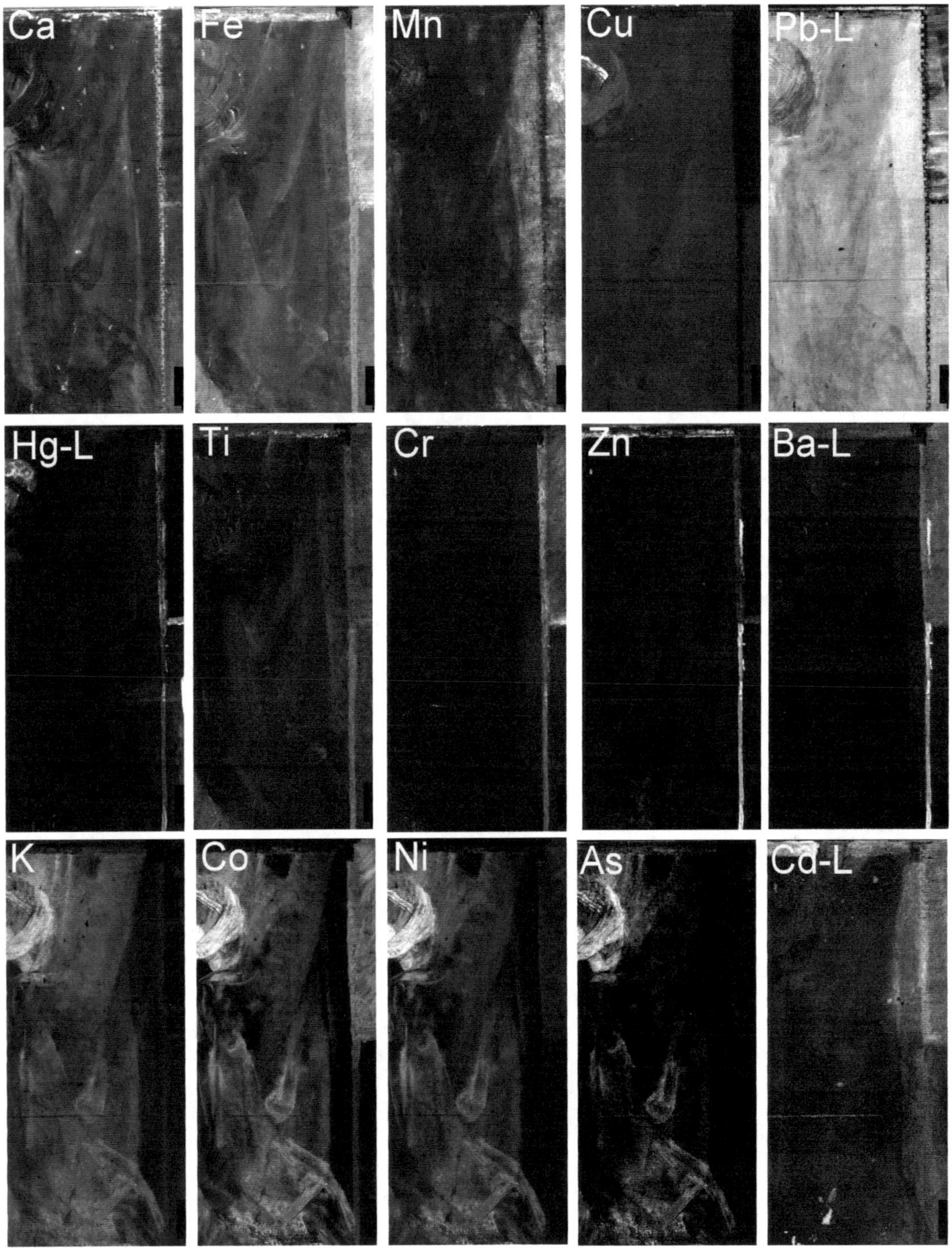

Figure 4. Elemental distribution images of the central part of the painting (approx. 60x120 cm^2). The elements were detected by their K or L level fluorescence radiation, as indicated in the labels.

Journal of Physics: Conference Series **499** (2014) 012013 doi:10.1088/1742-6596/499/1/012013

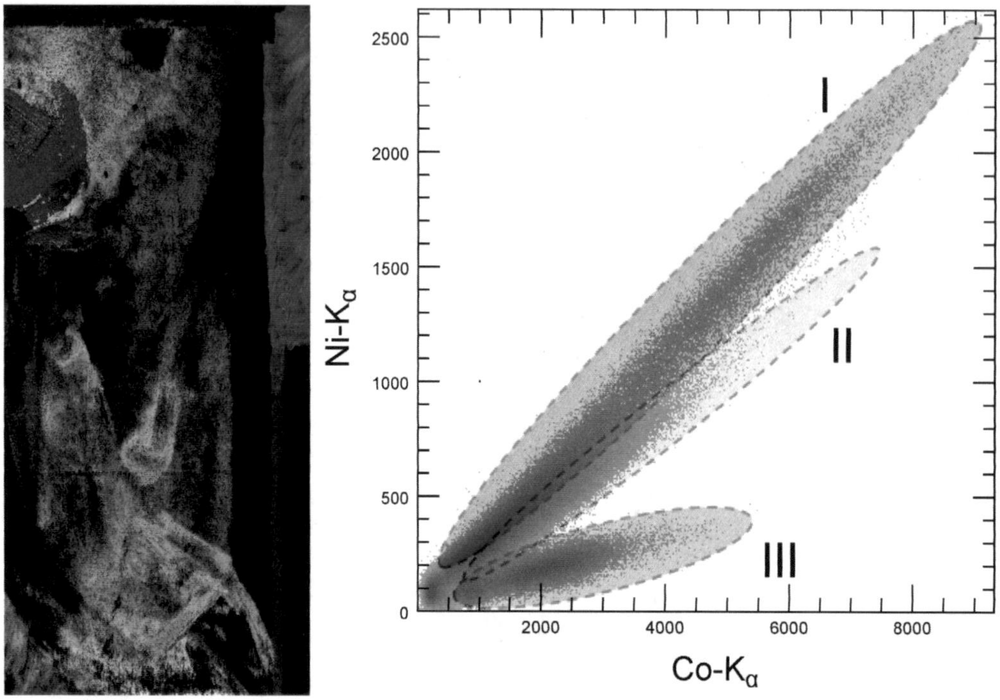

Figure 5. Co map colorized based on the Co/Ni ratio to visualize the extent of smalt (groups I and II) and drier (group III).

is present as a drier or as a pigment, but as Cd salts were not used until the 1840s they are, as aforementioned elements, good tracers for later restoration treatments and indicate heavy reworking of the area along the join.

K, Co, Ni and As are mainly present in the blue pigment smalt. However, from paint cross sections investigated by SEM-EDX it is known that Co and Ni are also present in a second compound: Co salts added as drier to paint in the 19th century. These two compounds cannot be differentiated by the Co signal alone. However, since Co ores also contain Ni the compounds synthesized from these minerals feature a characteristic Co/Ni ratio.

In the Co/Ni scatter plot in Figure 5, three different groups can be distinguished: group I and II represent smalt, while group III represents the drier. The fact that the smalt signal is distributed over two groups may suggest that two different kinds of smalt were used in the creation of the painting. It is obvious that the statistics of the scatter plot are insufficient to fully resolve the groups I and II.

However, the Co associated with smalt and that used as a drier (group III) can be clearly distinguished; the distribution of these Co containing compounds can be visualized by coloring the Co distribution image in Figure 5 accordingly. The smalt distribution image reveals the folds of the original curtain and shows that under the surface a considerable amount of the original paint is still present, except in the area directly along the join, where much of the original paint was removed.

4.2. NMF

The study of scatter plots can allow for the distinction between compounds of similar elemental composition. However, this requires a high degree of user intervention. Especially, if not a single

Journal of Physics: Conference Series **499** (2014) 012013 doi:10.1088/1742-6596/499/1/012013

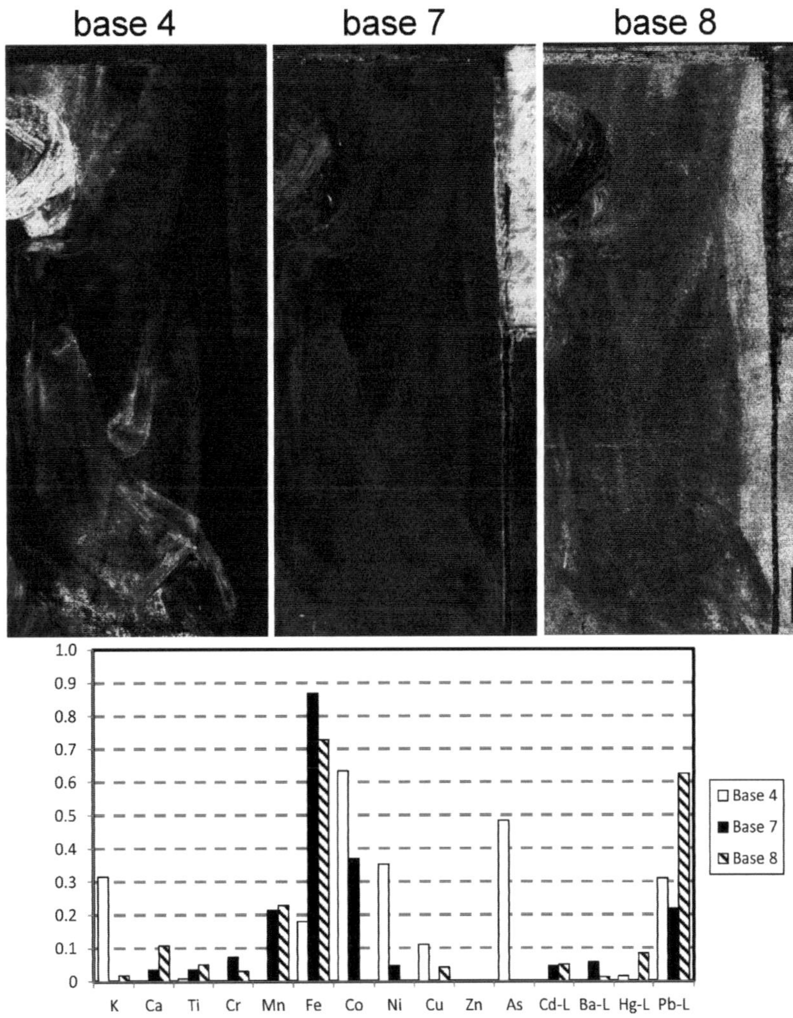

Figure 6. Three bases obtained by NMF. Base 4 represents smalt, base 7 the Co salt driers, while base 8 highlights the area in that no original paint is left. The scale of the bar plot is normalized intensity.

data set, such as in this case, but multiple from a series of measurements are studied or the correlation of several elements is of interest. In the case of *Saul and David* several regions of heavy reworking were known from the XRR, so that it was suspected from the beginning that the Co at the joins was not present in the pigment smalt but a different compound, which was confirmed by the SEM-EDX investigation of paint samples. In more complex or subtle cases this might have gone unnoticed.

The NMF[2] was applied on the stack of 15 elemental distribution images, each featuring 495x1074 pixels, shown in Figure 4. Using processed XRF data allowed to reduce the volume of data and so allowed for fast processing. Only signals containing information on the paint layers of the painting were considered; Ar signals, which originate from the air, were not included, as

[2] We used the NMF implementation of the Python Matrix Factorization Module (pymf [12]).

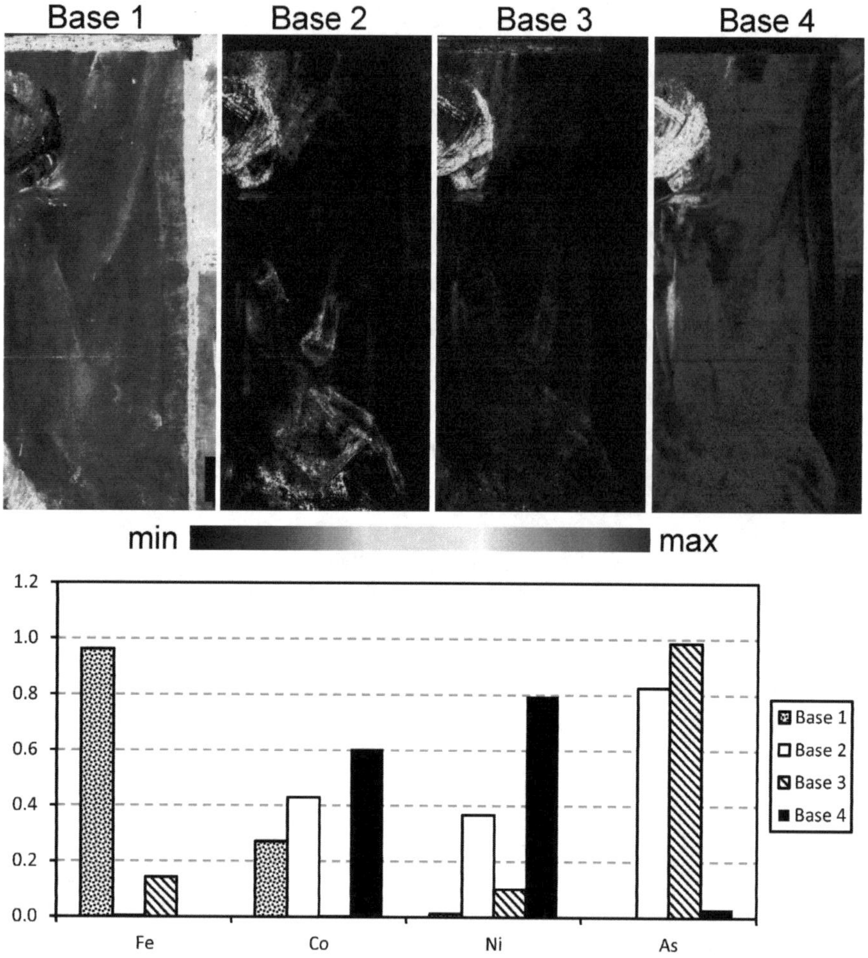

Figure 7. Four bases obtained by NMF on the elemental distribution images of Fe, Co, Ni and As. The bases 1, 2 and 4 are equivalent to the groups I, II and III in Figure 5. Base 3 illustrates the heterogeneous distribution of the As signal.

was the intensity of scattered primary radiation, which contains strong contributions from the wooden stretcher behind the painting. In case of multiple fluorescence lines of the same element only the most intense one was taken into account.

For the NMF investigation the data was scaled by dividing each distribution image by the square root of its mean value. This scaling step can be considered as a way of weighing the contribution of each image using the uncertainty of the average intensity as weight [1]. The results shown were obtained with 8 bases, in order to reduce the dimensionality of the data set by a factor of two. The calculation was aborted after 100 iterations, which was found sufficient for the decomposition of the data set. These 100 iterations took ca. 35 seconds on a standard Intel QuadCore CPU with 3.3 Ghz and 8 GB installed memory. As NMF starts with a random set of bases to iteratively improve the degree to that the variance of the data set is explained, the chosen seed for the random number generator has an considerable influence on the results. So, the processing in NMF was repeated 10 times with 100 iterations each and different seeds but otherwise identical parameters.

In Figure 6 three of the bases obtained are shown. Base 4 represents smalt and contains next to Co the elements K, Fe, Ni and As. These are typical minor components of smalt. Also Pb, which is virtually omnipresent in the painting, can be found. Base 7 represents the Co salt drier and contains next to Co and low amounts of Ni a considerable amount of Fe. Base 8 highlights the area of paint loss and contains the elements of the pigments used to fill the gap. Mainly Mn from dark earth pigments and Ca and Cd. The enhanced contribution of Pb to this base is assumed to be the result of a thinner covering layer, absorbing less of the Pb-L radiation emitted from the ground layer(s). The Co in the join is represented in another base (not shown). Given the heterogeneous distribution of material in the join area, no accurate representation of it was obtained. In all 10 calculations using different seeds the discrimination between smalt and drier was possible. The identification of the paint loss zone (base 8) was possible in 9 of 10 calculations. In the 100 iterations used no consistency between **W** and **H** was achieved for the different seeds, which was not the aim of the decomposition, but a correct decomposition of different components.

In the NMF of the whole data set no distinction between the different smalt groups (I and II in Figure 5) was obvious, albeit some bases indicated a separation between different groups of smalt. The *a priori* knowledge of two smalt groups could be reproduced, by limiting the NMF to the elemental distribution images of Fe, Co, Ni and As and increasing the number of iterations to 500 with four bases (see Figure 7). K, which is also a minor component of smalt, was omitted in this approach, as it is considerably lighter than the other elements and suffers stronger from absorption artifacts. The data was binned to 247x537 pixels to accelerate processing and normalized to its mean value. This results in all four elemental distributions having the same weight, independent of uncertainty and sensitivity of the instrument.

Base 1 represents earth pigments and the Co salts used as drier and are equivalent to group I in Figure 5. Base 2 is the equivalent to group II, while base 4 is equivalent to group III, which featured the highest Ni content relative to Co. Base 3 shows something not apparent in the scatter plot: the distribution of As and Co do not completely agree, but in some regions As is enhanced. If this difference is due to compositional variations or absorption effects has to be investigated further. It is a curious oddity that in base 3 no Co is present, albeit its distribution correlates with that of As in the elemental distribution images and in many other NMF treatments both elements were found in the same base.

While the elements in Figure 7 were chosen based on *a priori* knowledge to reproduce the manual results, it is quite reasonable that a researcher, experienced with the composition of smalt and XRF might select these elements. For an automatic separation of two smalt groups, improved NMF approaches such as those mentioned above might be helpful [5–7].

5. Conclusions

We have successfully shown in a case study how the NMF function of the pymf module supports the interpretation of a complex data set with a moderate degree of redundancy (Co, Ni, K and As being all present in smalt) and compounds with similar elemental composition (smalt and Co salts added as drier, both containing Co and Ni). A correct decomposition of components of the data set was possible with only minimal amounts of *a priori* knowledge.

Albeit the bases obtained suggested the presence of more than one smalt component, it could only be visualized by using *a priori* knowledge from scatter plots of the Co/Ni ratio. It is expected that NMF with more elaborate selection rules, considering e.g. sparseness or convexity constraints, would considerably improve the bases obtained.

While the manual interpretation of the scatter plot allowed for the discrimination between three Co containing compounds, it was rather time consuming. Further, it was also limited to the correlation of two elements and so the chances for overlooking the presence of Co in more than one component were present. An application of NMF at that moment would have

Journal of Physics: Conference Series **499** (2014) 012013 doi:10.1088/1742-6596/499/1/012013

simplified the interpretation of the data. In the original study the distinction between original and later added paint made it possible for the conservator to make a more accurate description of the painting's true condition and an informed conservation treatment proposal taking this new information into account.

We expect that in the future NMF and other routines of the pymf module with an easy to use graphical user interface will play a role in the interpretation of XRF data, as might NMF routines from other software packages, since an application of NMF can considerably accelerate the interpretation of the data.

Acknowledgments

The authors would like to thank Petria Noble of the Mauritshuis Museum (NL), Koen Janssens of the University of Antwerp (BE) and Joris Dik of the Delft University of Technology (NL) for their support in the acquisition of the original data set and their interpretation. The German Federal Ministry of Education and Research (BMBF) is acknowledged for the financial support (Verbundprojekt 05K2012).

References

[1] Vekemans B, Janssens K, Vincze L, Aerts A, Adams F and Hertogen J 1997 *X-Ray Spectrom.* **26** 333–346

[2] Kotula P G, Keenan M R and Michael J R 2003 *Microsc. Microanal.* **9** 1–17

[3] Friel J J and Lyman C E 2006 *Microsc. Microanal.* **12** 2–25

[4] Lee D D and Seung H S 1999 *Nature* **401** 788–791

[5] Ding C, Li T and Jordan M 2010 *IEEE Trans. Pattern Anal. Mach. Intell.* **32** 45–55

[6] Cutler A and Breiman L 1994 *Technometrics* **36** 338–347

[7] Thurau C, Kersting K, Wahabzada M and Bauckhage C 2012 *Data Min. Knowl. Disc.* **24** 325–354

[8] Noble P, van Loon A, Alfeld M, Janssens K and Dik J 2012 *Techné* **35** 36–45

[9] Alfeld M 2013 *Development of Scanning macro-XRF for the Investigation of Historical Paintings* Ph.D. thesis University of Antwerp, Belgium, and University of Hamburg, Germany, p. 199–202

[10] Alfeld M, Janssens K, Dik J, De Nolf W and Van der Snickt G 2011 *J. Anal. At. Spectrom.* **26** 899–909

[11] Vekemans B, Janssens K, Vincze L, Adams F and Van Espen P 1994 *X-Ray Spectrom.* **23** 278–285

[12] Pymf: Python matrix factorization module, `https://code.google.com/p/pymf/`, last visited 02.12.13

Journal of Physics: Conference Series **499** (2014) 012014 doi:10.1088/1742-6596/499/1/012014

Emerging Patterns in the Distribution of Trace Elements in Ovarian, Invasive and In-Situ Breast Cancer

A Al-Ebraheem[1], E Dao[1], K Geraki[2], and M J Farquharson[1]
[1] *Department of Medical Physics and Applied Radiation Sciences, McMaster University, 1280Main St W Hamilton, Ontario, L8S 4K1, Canada*
[2] *Diamond Light Source, Harwell Science and Innovation Campus, Didcot, Oxon, OX11 0DE,UK*
Email: farquhm@mcmaster.ca

Abstract: Breast cancer is the most common cancer and ovarian cancer is the 8th most common cancer affecting women worldwide. This study highlights the changes of trace element levels accompanied by the progression from ductal carcinoma in situ (DCIS) to invasive ductal carcinoma (IDC) of the breast, using micro probe Synchrotron Radiation X-ray Fluorescence (μSRXRF). The average values for the increase in Ca, Fe and Zn in tumour regions with respect to surrounding regions for the DCIS samples were significantly higher compared to the increase in the IDC samples (P <0.01).This study was also carried out to find a connection between ovarian cancer and breast cancer with respect to the cellular distribution of Ca, Cu, Fe, and Zn. For IDC, DCIS and ovarian cases, the statistical analysis reveals a significant increase in the levels of Ca, Cu and Zn concentrations in cancer tissue when compared to the normal surrounding tissue. For Fe, the differences between tumour regions with respect to surrounding regions were found to be not significant in IDC and ovarian cases. In DCIS cases, the results reveal a significant increase in the levels of Fe in cancer tissue when compared to the surrounding normal breast tissue (P <0.01).

Introduction

The most common type of breast cancer for women is mammary ductal carcinoma and it presents in two forms: invasive ductal carcinoma (IDC), an infiltrating, malignant and abnormal proliferation of neoplastic cells in the breast tissue; or ductal carcinoma in situ (DCIS), a noninvasive, potentially malignant cancer, where abnormal cells are restricted to the ducts. About 50% of untreated DCIS lesions are believed to rapidly transit to IDC, with a high degree of variability in this progression [1]. Our work to date has investigated the cellular distribution of Ca, Cu, Fe and Zn in IDC of breast tissue using micro probe Synchrotron Radiation X-ray Fluorescence (μSRXRF) [2-5]. These studies reveal an average increase for Ca, Cu and Zn levels in areas of malignancy. The levels of Fe on average showed an increase but in some samples there was a decrease in the tumour regions. The first part of the study reported here will focus on the investigation to determine whether the progression from DCIS to IDC is accompanied by changes of trace element levels.

Ovarian cancer is the 8th most common cancer affecting women in the world [6]. Rosen DG *et al.* reported that ovarian cancer is responsible for 140,000 deaths around the world each year and has the lowest survival rate of all female cancers. Only 45% of patients are likely to survive after 5 years of diagnosis [7]. The most recent data regarding ovarian cancer reports that in 2008, there were 12.7 million new cancer cases; 225,000 of which were ovarian cancer [6]. Ovarian cancer is often diagnosed as a secondary cancer alongside an initial primary cancer. Cases have been reported where ovarian cancer is diagnosed in addition to breast cancer and it has been found that women inflicted with breast cancer or with family history of breast cancer are also at an increased risk of developing ovarian cancer. Consequently with ovarian cancer being so closely linked to breast cancer, the second part of the study reported here was carried out to find a connection between ovarian cancer and breast cancer with respect to the cellular distribution of Ca, Cu, Fe, and Zn using μSRXRF. Finding a connection between the two diseases may provide insight with regards to many unknown factors of the two diseases as it has been found that breast and ovarian cancer tissues have similar properties, and thus the same biomarkers are able to be used to diagnose and monitor the disease progression in both cancers [8]. Utilising in vivo experiments and the use of chelators in mice, Merlot, *et al.* found that a slight decrease in Fe levels could be observed in the cancer cells [9]. It was also found that Cu levels typically tend to be elevated in both breast and ovarian cancer tissues. This trend is seen in different age groups and geographical locations of those inflicted with the diseases. The levels of Zn were also investigated and found to remain constant.

Samples:

The IDC, DCIS and ovarian samples were obtained from the Cancer Research UK Tumour Pathology Group, Department of Clinical Laboratory Sciences, John Radcliffe Hospital, University of Oxford. All samples were Formalin Fixed Paraffin Embedded (FFPE) tissues arranged as micro arrays of 1.0 mm diameter and 10μm thickness. For each tissue micro array (TMA), two slices were cut from the paraffin block, one being 10 μm thick, the other being 5 μm thick and cut adjacent to the 10 μm slice. The 5 μm thick slice was mounted on a standard glass slide and then stained using Hematoxylin and Eosin stain (H&E) staining. The H&E slide was then optically imaged to produce high resolution images which clearly identify tumour regions in the samples and can be used as a reference slide for the elemental distribution maps produced from the experimental slide. The 10 μm thick slide was mounted on 4 μm Ultralene XRF film that was stretched and held in a bespoke frame.

Method:

The IDC, DCIS and ovarian sample measurements were carried out using the Synchrotron Radiation X-ray Fluorescence (SRXRF) at beamline I18, DIAMOND, UK and FLUO beamline at ANKA, Karlsruhe, Germany**.** Figure 1 shows the sample micro-array slide and a schematic of the experimental set-up. Figure 1 also shows the 90° geometry between the incident beam and the detector, which is maintained to utilise the strong polarization of the synchrotron photon beam. This arrangement provides significant suppression of the scattered incident photons in the plane of the electron orbit, while the fluorescence signal which is required for the measurement remains unaffected. The sample position is set at 45° to the incident photon beam. The experimental set up took on a similar form at the two facilities with specific details at each Beamline as given in table 1. The data collected at each point in the scan is in the form of an elemental spectrum that is collected using a Si (Li) detector (Oxford Instruments) at the FLUO beamline and a 4-element Vortex SDD detector at beamline I18. These spectra were analysed using AXIL and PyMCA programs respectively both of which are designed to fit the XRF peaks, remove the background and separate the overlapping peaks. The final metal distribution for each metal is composed by the element's peak intensity in each sample position. Maps of the Zn, Fe, Cu and Ca distributions were produced using IDL v6.3 (ITT Virtual Information Software). This program enabled regions of interest to be set on the maps encompassing tumour cell regions and regions of surrounding non-tumour tissue which is useful for further statistical analysis. Statistics obtained from these regions including mean pixel value and the standard error of the means. The parameter compared is the mean of Ca, Fe, Cu and Ca levels in tumour (T) and normal (N) regions. When comparing the changes of trace element levels between IDC and DCIS cases, the parameter compared is the average value for the percentage difference in the metal levels in tumour regions with respect to surrounding normal regions for each case (i.e for each element ((Tumour content-Normal content)/Normal content)*100%. The Shapiro–Wilk test was used for each element in each category to test if the data set was consistent with a Gaussian distribution function. For IDC, DCIS and ovarian tumour and normal data, Wilcoxon Rank test was used for comparing the difference in Ca, Fe, Cu and Zn levels for non normally distributed related data, while paired t-test was used for normally distributed related data. For comparing the percentage difference in Ca, Fe, Cu and Zn levels in IDC samples with respect to DCIS samples, *P* values from Mann Whitney U and Independent t test were obtained.

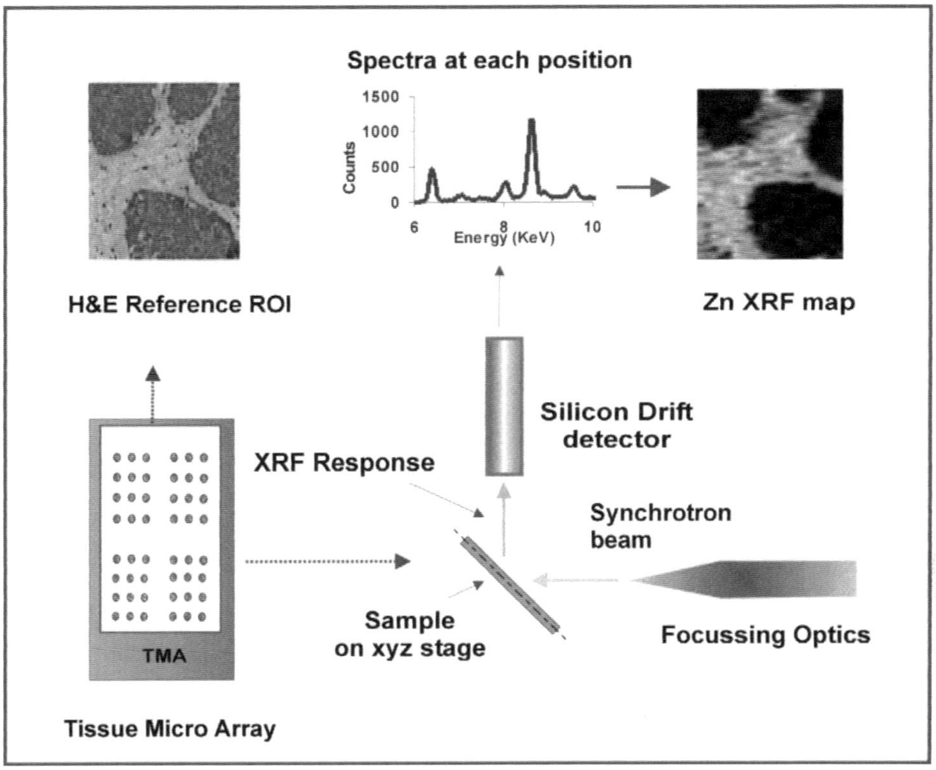

Figure 1: Schematic diagram showing samples arranged as a tissue micro array and mounted on a sample holder held on an xyz translator stage. The sample array is positioned at 45° to the synchrotron beam.

Beam line	FLUO at ANKA	I18 at DIAMOND
Incident Energy	10.1 keV	11 keV
Beam Size	5 μm x 15 μm	5 μm x 5 μm
Time/Pixel	Varied	4 seconds
Detector	Si(Li) detector (Oxford Instruments)	4-element Vortex detector
Fitting software	AXIL program	PyMCA program
Final parameter compared	Total metal Kα counts	Total metal Kα counts normalised to the incident scatter peak area

Table 1: μSRXRF beamlines experimental set up details.

Result and Discussion:

1- Elemental Mapping :

DCIS, IDC and ovarian carcinoma samples were scanned using a μSRXRF technique. Figure 2 shows an example of the elemental distributions maps of Zn, Fe, Cu, and Ca in two samples of DCIS tissue, two samples of IDC tissue and two ovarian tissue samples. The dark areas of the H&E stained reference image show the tumour areas, and the lighter regions are the normal tissue. Correlations were found between the stained reference images and the elemental distributions indicating an increase in element concentrations in most of the tumour regions of the sample (darker coloured areas correspond to higher metal content), but it can also be seen that the elemental distribution of Fe in some samples displays the reverse behaviour within the tumour areas with respect to the normal.

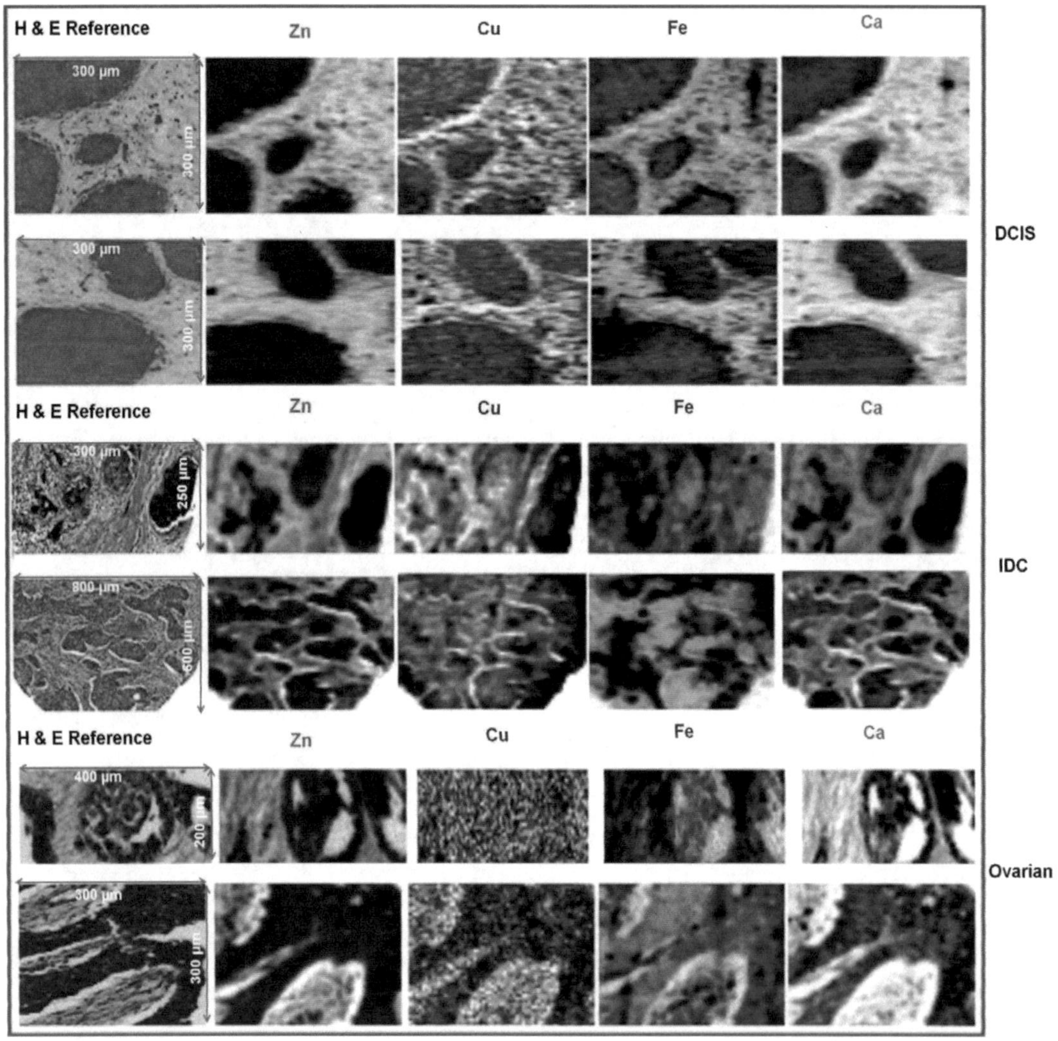

Figure 2: Examples of sample maps, showing the reference H&E stained reference image with the corresponding Ca, Zn, Cu and Fe distribution maps. Examples are shown of DCIS, IDC and ovarian carcinoma.

2- Elemental concentration in IDC and DCIS samples:

The relative levels of Zn, Cu, Fe and Ca were estimated in 20 IDC and 20 DCIS FFPE samples and normal surrounding breast tissue. Table 2 shows a summary of the descriptive statistical parameters including the mean, the standard error of the means and the *P* values for comparing the difference in Ca, Fe, Cu and Zn levels in tumour regions (T) with respect to surrounding normal regions (N) for DCIS and IDC samples.

Groups	Mean (Counts) (n=20)	Std. Error Mean (Counts)	*P* values
IDC-Ca-T	2144.73	420.78	<0.01*
IDC-Ca-N	1083.15	208.05	
IDC-Fe-T	140.61	33.45	0.84
IDC-Fe-N	129.60	26.14	
IDC-Cu-T	51.07	17.41	<0.01*
IDC-Cu-N	38.06	11.97	
IDC-Zn-T	282.84	75.62	<0.01*
IDC-Zn-N	109.83	22.79	
DCIS-Ca-T	3223.68	180.55	<0.01*
DCIS-Ca-N	1050.95	79.48	
DCIS-Fe-T	96.30	10.57	<0.01*
DCIS-Fe-N	38.12	4.98	
DCIS-Cu-T	80.70	7.95	<0.01*
DCIS-Cu-N	57.76	5.00	
DCIS-Zn-T	201.75	12.61	<0.01*
DCIS-Zn-N	41.39	2.81	

Table 2: Statistical summary of the mean and the standard error of the means of Ca, Fe, Cu and Ca levels in tumour (T) regions with respect to the normal (N) regions in DCIS and IDC sample. The corresponding *P* values from Wilcoxon Rank test (2-tailed).*Significantly different at $P < 0.01$.

The statistical analysis reveals an average increase in the levels of Ca, Cu and Zn in IDC tissue when compared to the normal surrounding breast tissue ($P < 0.01$). For Fe, the differences between tumour regions with respect to surrounding regions were found to be not significant in IDC cases ($P = 0.84$). In DCIS cases, the results reveal a significant increase in the levels of Ca, Fe, Cu and Zn levels in DCIS tissue when compared to the surrounding normal breast tissue($P < 0.01$). Table 3 shows a summary of the descriptive statistical parameters including the percentage difference of the mean levels in tumour regions with respect to surrounding normal regions for IDC and DCIS samples (i.e for each element ((Tumour content-Normal content)/Normal content)*100%, the standard error of the percentage difference of the means, and the *P* values for comparing the percentage difference in Ca, Fe, Cu and Zn levels in IDC samples with respect DCIS samples.

Groups	% Difference of the means (Counts) (n=20)	Std. Error Mean (Counts)	*P* values
IDC-Ca	91.42	9.99	<0.01*[2]
DCIS-Ca	229.80	24.91	
IDC-Fe	19.29	15.13	<0.01*[1]
DCIS-Fe	178.32	37.61	
IDC-Cu	33.84	5.38	0.397[2]
DCIS-Cu	40.96	6.31	
IDC-Zn	127.75	17.80	<0.01*[1]
DCIS-Zn	419.65	44.44	

Table 3: Statistical summary of the percentage difference of Ca, Fe, Cu and Ca levels in tumour regions with respect to the normal regions for IDC and DCIS samples, the standard error of the percentage difference of the means, and the *P* values for comparing the percentage difference in the mean metal levels in IDC samples with respect DCIS samples. The corresponding *P* values from Mann Whitney U[1] (2-tailed) Independent t test[2] (2-tailed). *Significantly different at $P < 0.01$.

The null hypothesis in this part of the study is that: the percentage differences in the metal levels in the two groups of cancer (IDC and DCIS) are equal. The null hypothesis is rejected when the p-value is less than 0.05. This test will enable us to assess if the progression from DCIS to IDC of the breast is accompanied by significant metal changes. The average values for the percentage increase in Ca, Fe and Zn in tumour regions with respect to surrounding regions for the DCIS samples were significantly higher compared to the percentage increase in the IDC samples ($P < 0.01$). The percentage difference between tumour and normal regions of Cu is not significantly higher in DCIS cases compared to the IDC cases. The results presented in our most recent publication [5] that combined data from 128 formalin fixed paraffin embedded IDC samples indicate the importance of correlating the changes in elemental levels to the clinical prognostic factors such as oestrogen receptor (ER) status. In order to take this into account the DCIS and IDC samples were classified on the basis of their ER status. Table 4 shows a summary of the results including only the ER +ve group DCIS and IDC sample. The clinical data was not available for all the cases, therefore a comparison of the ER negative (-ve) group has not been included.

Groups ER +ve	% Difference of the means (Counts) (n=6)	Std. Error Mean (Counts)	*P* values
IDC-Ca	114	18	0.14
DCIS-Ca	203	72	
IDC-Fe	-9	11	0.033*
DCIS-Fe	268	119	
IDC-Cu	30	8	0.36
DCIS-Cu	56	26	
IDC-Zn	166	43	0.04*
DCIS-Zn	337	77	

Table 4: Statistical summary of the percentage difference of Ca, Fe, Cu and Ca levels in tumour regions with respect to the normal regions for IDC and DCIS ER +ve samples, the standard error of the percentage difference of the means, and the *P* values for comparing the percentage difference in the mean metal levels in IDC samples with respect DCIS samples. The corresponding *P* values from Independent t test (2-tailed). *Significantly different at $P < 0.05$.

For the ER +ve DCIS and IDC samples, the statistical analysis reveals that the percentage difference in Fe and Zn in tumour regions with respect to surrounding regions for the DCIS samples were significantly higher compared to the percentage in the IDC samples.

3- Elemental concentration in ovarian cancer samples:

This part of the study examined the cellular distribution of the trace of the same four elements in 20 FFPE ovarian cancer samples and normal surrounding ovarian tissue. Table 5 shows a summary of the descriptive statistical parameters including the mean values and the *P* values for of the comparisons.

Groups Tumor (T) and Normal (N) ovarian tissue	Mean Normalised Counts (n=20)	Std. Error Mean Normalised Counts	*P* values from
Ca-T	0.026434	0.00226	<0.01[1]*
Ca-N	0.011694	0.00105	
Fe-T	0.004483	0.00024	0.14[1]
Fe-N	0.004053	0.00017	
Cu-T	0.000651	0.00002	<0.01[1]*
Cu-N	0.000614	0.00002	
Zn-T	0.006005	0.00038	<0.012*
Zn-N	0.003191	0.00021	

Table 5: Statistical summary of the mean and the standard error of the means of Ca, Fe, Cu and Ca levels in tumour (T) regions and the normal (N) regions of the ovarian samples. The corresponding *P* values from paired T test[1] (2-tailed) or Wilcoxon test[2] (2-tailed).*Significantly different at $P < 0.01$.

The null hypothesis in this part of the study is that: the trace element levels in tumour and normal ovarian tissues are equal. When the p-value is less than 0.05, the null hypothesis is rejected. The statistical analysis reveals a significant increase in the levels of Ca, Cu and Zn concentrations in ovarian cancer tissue when compared to the normal surrounding ovarian tissue. For Fe, the differences in between tumour regions with respect to surrounding regions were found to be not significant.

Conclusion:

In the first part of the study, the analysis reveals a significantly higher percentage difference in Ca, Fe and Zn in tumour regions with respect to surrounding regions for the DCIS samples compared to the percentage difference in the IDC samples. When the samples were reclassified in terms of their ER +ve status the statistically significant differences between the two types of breast tumour are again observed for Fe and Zn. The Fe result is in agreement with our previous study [5] which showed that imbalance in Fe concentration (deficiency) should be viewed as an important risk factor that is associated with aggressive features of the cancer. The second part of the study shows that the trend of the differences in the levels of Ca, Fe, Cu and Zn elements is similar between the IDC and ovarian cancers. For IDC and ovarian cases, the statistical analysis reveals a significant increase in the levels of Ca, Cu and Zn concentrations in cancer tissue when compared to the normal surrounding tissue. For Fe, the differences between tumour regions with respect to surrounding regions were found to be not significant in IDC and ovarian cases. This research will serve as a base for future studies aimed at exploring advanced cancer treatment options such as chelators that target metal concentrations.

Acknowledgements

This work was carried out with the support of the Diamond Light Source, UK and the FLUO beamline at ANKA, Germany. We are grateful to the beamline teams of F. Mosselmans, and R. Simon for their invaluable assistance and support. Tissue samples were provided by Prof A.L. Harris (Cancer Research UK Molecular Oncology Laboratories, Weatherall Institute of Molecular Medicine,

University of Oxford) and Dr. R. Leek (Cancer Research UK Tumour Pathology Group, Nuffield Department of Clinical Laboratory Sciences, University of Oxford).

References:

[1] Collins L C et al 2005 *Cancer* **103** 1778-1784
[2] Al-Ebraheem A et al 2010 *X-Ray Spectrometry* **39** 332-337
[3] Farquharson M J et al 2009 *Physics in Medicine and Biology* **54** 4213-4223
[4] Geraki K et al 2007 X-*Ray Spectrometry* **37** 12-20
[5] Al-Ebraheem A et al 2013 *X-Ray Spectrometry* **42** 330-336
[6] Ferlay J et al 2008 Incidence and Mortality Worldwide: IARC CancerBase *GLOBOCAN* **2** No. 10
[7] Rosen D G et al 2010 *Front Biosci* **14** 2089-2102
[8] Ullah M F and Aatif M 2009 *Cancer Treatment Reviews* **35** 193-200
[9] Merlot A M et al 2013, *Antioxidants & Redox Signaling* **18** 973-1006

Journal of Physics: Conference Series **499** (2014) 012015 doi:10.1088/1742-6596/499/1/012015

Influence of depositional environment in fossil teeth: a micro-XRF and XAFS study.

I M Zougrou[1], M Katsikini[1*], F Pinakidou[1], L Papadopoulou[2], E Tsoukala[2] and E C Paloura[1]

[1] School of Physics, Aristotle University of Thessaloniki, 54124 Thessaloniki, Greece.
[2] School of Geology, Aristotle University, 54124 Thessaloniki, Greece.

*E-mail: katsiki@auth.gr

Abstract. The formation of metal-rich phases during the fossilization of vertebrate fossil teeth, recovered from various deposition environments in northern Greece, is studied by means of synchrotron radiation X-ray fluorescence (SR-XRF) as well as Fe and Mn *K* edge X-ray absorption fine structure (XAFS) spectroscopy. XRF line-scans from the samples' cross-sections revealed different contamination paths for Mn and Fe. The two-dimensional XRF maps illustrate the spatial distribution of P, Ca, Mn and Fe as well as the precipitation of Fe-rich phases in cementum, dentin and dentinal tubules. Goethite, lepidocrocite and ferrihydrite were detected in the samples' cross-section by means of Fe *K* edge EXAFS spectroscopy. Moreover the Fe and Mn *K* edge EXAFS revealed the presence of vivianite and birnessite (MnO_2) on the external surface of two samples.

1. Introduction

The study of fossil teeth composition and chemical alterations is of great importance not only for palaeoenvironment reconstruction and identification of the fossilization conditions but for conservation and restoration purposes as well. Mammal teeth consist typically of a crown composed of dentin covered by an enamel layer and a cementum layer which coats the tooth pulp. Although the inorganic component of all layers is carbonate hydroxylapatite (HAp), the microstructure and HAp percentage vary among the layers. More specifically, enamel consists of 97% wt. HAp crystals while in dentin and cement the corresponding component is 70% wt. Amelogenin proteins comprise the main organic component of enamel while dentin and cement comprise of type I collagen [1,2]. Contrary to other mammals, the structure of the occlusal surface in the cheek teeth of hypsodont animals such as the *Hipparion*, is more complex. Enamel in the *Hipparion* molar and pre-molar teeth forms several folds in the coronal dentin, named infundibula, which result in the formation of enamel ridges elevated from the occlusal surface. Cementum fills the infundibula and covers the enamel (coronal cementum). The *Hipparion* cheek teeth have five cusps: the protocone, the protoloph, the metaloph, the paracone and the metacone (figure 1a). The interior of the two complex enamel structures forming plications in the centre of the tooth are called prefossette and postfossette [3,4]. Although the inorganic phases of enamel, dentin and cementum are preserved during fossilization, porosity and degradation of the organic content contribute to the chemical alterations of the tooth induced by diagenetic processes. Dentin is more porous than enamel, containing dentinal tubules of 1-3 μm diameter which may be enlarged by dissolution or filled with diagenetic minerals. Moreover, the collagen content of the dentin and cementum is more easily dissolved rendering these tissues more prone to diagenetic alterations than enamel [5].

In this study we investigate the diagenetic alterations of fossilized teeth of different animal taxa from three excavation sites in northern Greece. The samples were selected on the basis of the different dental tissue structure and different deposition environment. The contamination profiles of dental tissues were studied by micro-X-ray fluorescence (μ-XRF) line-scans across the cross-sections of the samples. Two-dimensional (2D) XRF maps were recorded to investigate the distribution of P, Ca, Mn and Fe in the dental tissue. The local bonding environment of Fe and Mn, which are the most common soil contaminants in fossil specimens, was determined by micro-X-ray absorption fine structure (μ-XAFS) spectroscopy. The bonding environment and oxidation state of Fe was assessed using the X-ray absorption near edge structure (XANES) portion of the Fe *K* edge XAFS spectra. The extended XAFS (EXAFS) part of the spectrum, in combination with XANES, provides information on the metal-rich phases. Color alterations at the external surface of two samples were also investigated by means of Mn and Fe *K* edge XAFS spectroscopy.

2. Materials and Methods

Fossil teeth of various geological ages recovered from three excavation sites in northern Greece were studied by means of μ-XRF and XAFS spectroscopies. The sample names and descriptions are listed in table 1 and corresponding photographs are shown in figure 1.

Table 1: Sample names and characteristics

Sample name	Animal taxum	Specimen	Excavation site	Age
MIL5 753	*Hipparion* sp.	Upper molar tooth	Milia, Grevena (site 5)	Earlymost Villafranchian (MNQ 16) Plio-Pleistocene
MIL5 742	*Sus arvernensis*	Canine tooth	Milia, Grevena (site 5)	MNQ 16
XEK 13	*Elephas antiquus*	Tusk fragment (Incisor tooth)	Xerias, Kavala	Middle Pleistocene
LAC 9141	*Ursus ingressus*	Canine tooth	Almopia Speleopark, Aridaia	42 ka BP (radiocarbon dating) Late Pleistocene

The Milia fossils were found in Pliocene - Quaternary deposits of the Grevena basin. The deposits consist of Pliocene lacustrine sediments, followed by a Plio-Pleistocene sequence of clastic deposits, mainly sands, gravels and conglomerates, with sandstone intercalation and silty lenses [6]. The fossil teeth were found at a depth of approximately 15 m and thus their preservation state was not influenced by seasonal temperature and humidity changes occurring at the surface [7]. The Xerias fossils were found during the construction of the Egnatia Highway. The Middle Pleistocene fossil tusk was preserved in red clay deposits and had evident deformation marks in the form of circular and radial cracks propagating from the outer surface to the centre of the tusks' cross-section, which were probably induced by the overlying sediment pressure [8]. Finally, the Almopia Speleopark is situated on the northern slope of the V-shaped Nikolaou Rema gorge, on the slopes of Voras Mountain. The gorge consists of Maastrichtian limestone of the Almopia zone. The bulk of the cavities of the Almopia Speleopark developed under phreatic conditions while neotectonic activity lifted the caves to the vadose zone. The caves are located at 460 – 560 m altitude while hot springs of 30 – 37.5 °C water temperature lay under at altitudes of 360 – 390 m [9]. The fossil bear tooth was excavated from the V3 square in the LAC I b Chamber of the Bear Cave at approximately 1.1 m depth. The Chamber of the cave was periodically flooded and the fossils were sporadically submerged in water [10].

The SR-XRF and μ-XAFS measurements were conducted at the BESSY-II Synchrotron Radiation Storage Ring of the Helmholtz Zentrum Berlin. The measurements were performed at the KMC-II

Journal of Physics: Conference Series **499** (2014) 012015 doi:10.1088/1742-6596/499/1/012015

beamline, at atmospheric pressure, using an energy dispersive fluorescence (XFlash-Rontec) silicon detector positioned on the horizontal plane at right angle to the incident beam. The total flux at the sample spot was in the range 10^9-10^{10} photons/sec [11]. The beam was focused using a circular capillary that reduced the beam size to 3.6 μm. The one dimensional (line-scans) and two-dimensional SR-XRF maps were recorded using a beam with excitation energy of 9.7 or 7.2 keV, impinging at 45° to the sample surface. For the construction of the XRF maps, the integrated count rate of the K_α emission peak of the corresponding element was taken into account [region-of-interest (ROI) integration]. The phosphor XRF maps were corrected for the Ca K_α escape peak. Macro-XAFS spectra were recorded at the same beamline after removing the capillary as well as at the C beamline of the DORIS storage ring at HASYLAB. In the latter case the beam size and photon flux on sample was 6 × 1 mm^2 and 10^9 photons/sec, respectively, and the spectra were recorded using a 7-element fluorescence detector. XAFS spectra were recorded at the Fe and Mn *K* edges provided that the metal concentration was sufficient. Porosity measurements are based on scanning electron microscopy - backscattering electrons (SEM-BSE) and optical microscopy images. For the porosity determination, the area of pores with diameter >7 μm existing in 500×500 μm^2 squares was taken into account.

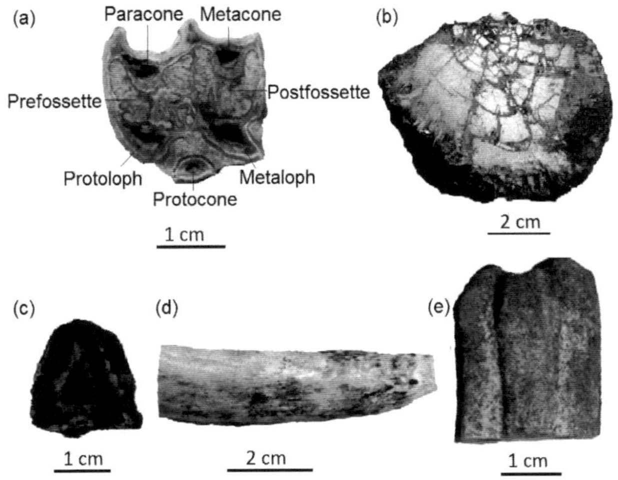

Figure 1 Photographs of the (a) *Hipparion* sp. upper molar tooth (MIL5-753) cross section. The five cusps as well as the pre- and postfossette are indicated, (b) *Elephas antiquus* tusk (XEK 13) cross section, (c) *Sus arvernensis* lower canine tooth (MIL5-742) cross section, (d) *Ursus ingressus* canine tooth (LAC 9141) lateral surface, (e) *Hipparion* sp. molar tooth (MIL5-753) lateral surface.

For the XRF and XAFS characterization the samples were cut into transverse sections and then they were cleaned, air-dried and ground with 400-800 SiC grit papers. The measurements were conducted on the cross sections and on the external untreated surfaces of the bulk samples. For the analysis of the EXAFS spectra the FEFF8 package was used [12,13]. The XANES spectra were subjected to linear background subtraction and normalization to the edge jump while the fitting of the isolated pre-edge peak was performed with the PeakFit software. Characterization results on fossil bones excavated at the site of Milia were published elsewhere [14].

3. Results and discussion

3.1. Synchrotron radiation X-ray fluorescence mapping

The contamination profiles and the related alterations in different dental regions of the *Hipparion* molar and the *Sus arvernensis* canine teeth were assessed using μ-XRF line-scans. A line-scan at the enamel – dentin region near the surface of the *Hipparion* molar and a second line-scan at the cementum region of the prefossette were recorded with excitation energy of 9.7 keV. The *Sus arvernensis* canine line-scan was recorded at a dentin region with excitation energy of 7.2 keV. The enamel surrounding the outer surface of the *Hipparion* tooth cusps is not well preserved and its thickness ranges from 0.5 to 1.5 mm. The line-scan of Ca, Mn and Fe K_α emission intensities are shown in figure 2. In order to compensate for the different emission and absorption probabilities of each element, the profiles were normalized according to the fundamental parameter method by $\varepsilon(E_{ijk})\omega_{ij}p_{ijk}\tau_{ij}(E_0)$ where ε is the detector's efficiency at the fluorescence energy E_{ijk} of the element i, ω_{ij} is the fluorescence yield, p_{ijk} is the transition probability of the kth line of the element i and τ_{ij} is the corresponding subshell photoelectric cross section at the energy of the impinging beam E_0 [14]. The profiles were not corrected for self-absorption effects due to the unknown matrix composition of the samples.

The distribution profile of Ca is related to the tooth bioapatite mineral, while Fe and Mn profiles might indicate incorporation in the bioapatite matrix and/or contamination from the burial environment. The profile of Ca is irregular in areas where the Fe concentration is high and fairly regular in areas where the Fe concentration is low. The decreasing Ca intensity in the *Hipparion* dentin region (figure 2a) extending to 1000-1200 μm length from the outer surface cannot be attributed to porosity. Based on SEM-BSE images the pore size diameter ranges from 7 to 70 ± 4 μm up to a distance of 1500 μm from the outer surface whereas in the same region the porosity ranges from 6.2 to $28.6\times10^{-4} \pm 0.2\times10^{-4}$. Thus, the variation in the Ca intensity is attributed to chemical changes in the matrix composition induced by dissolution and recrystallization processes rather than variations in the porosity values. In the *Hipparion* cementum region larger pores are observed (with size up to 98×56 μm^2) and the corresponding porosity increases up to $6.9\times10^{-3} \pm 0.2\times10^{-4}$. The local intensity minimum observed between x=430 and 530 μm in figure 2b is attributed to the presence of such pores. The Fe profile is highly irregular in both near surface dentin regions (figure 2a and c) and in the cementum region in the centre of the *Hipparion* tooth (figure 2b) showing various local maxima. The observed contamination paths of Fe may have resulted from diffusion and/or groundwater infiltration processes. The presence of Mn is minimal near the surface however a Mn maximum is observed in the *Hipparion* cementum revealing that Mn contamination originated from infiltration processes through the empty *Hipparion* cusps. This observation is also confirmed by SEM-BSE images which reveal that the metal-rich phases mainly precipitate in the pre- and post- fossette regions.

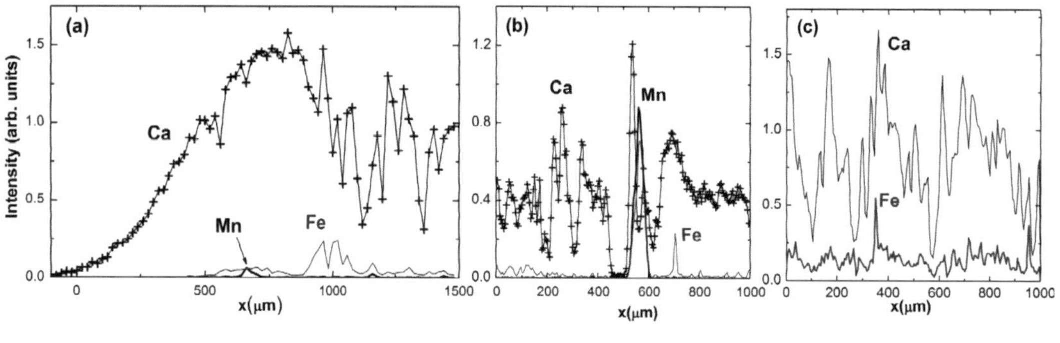

Figure 2. μ-XRF line-scans of Ca, Mn and Fe recorded at the cross sections of the (a) enamel-dentin region of the *Hipparion* molar tooth, (b) cementum region in the centre of the *Hipparion* molar tooth, (c) dentin region of the *Sus arvernensis* canine.

The Ca profile of the *Sus arvernensis* dentin region (figure 2c) is highly irregular revealing consecutive local variations in the apatite matrix. More specifically, local minima in the Ca concentration with a width of 25-85 μm, are observed. The Fe profile is also irregular but not directly correlated to the Ca profile. Local maxima in the Fe concentration, that denote Fe containing mineral precipitations, are observed both in spots of high and low Ca intensity. Reflected light optical microscopy images obtained in order to complement the μ-XRF line-scans, revealed three characteristic alterations in the *Sus arvernensis* dentin: a) the presence of secondary minerals in the dentin region next to the Enamel-Dentin Junction (EDJ) due to diagenetic cracks traversing the enamel, b) partially filled circular voids with diameter ranging from 1.6 to 12.5 ± 0.8 μm in a dentin area up to a distance of 63±4 μm from the EDJ, c) extensive microbial borings filled with secondary minerals extending up to 1000±4 μm from the EDJ. The borings replace most of the apatite matrix up to approximately 450 μm from the EDJ while their density is significantly reduced at larger distances from the EDJ. This altered area is macroscopically distinguished by its yellow color. The size of the borings in the yellow area ranges from 9.4×10.9 ± 0.8 μm^2 to 14×125 ± 0.8 μm^2 whereas the size of the dentinal tubules ranges from 1.6 to 6.3 ± 0.8 μm. Thus, the variation in the Ca intensity in the linear scan reflects the replacement of apatite by secondary minerals that fill the microbial borings.

The two-dimensional maps recorded at the dentin region of the *Hipparion* and the *Sus* teeth are shown in figures 3 and 4, respectively. The Ca and P maps reveal a similar spatial distribution of these elements indicating their co-participation in the formation of tooth apatite. A diagenetic crack observed in the *Sus* tooth is indicated by an arrow (figure 4). Metal containing, mainly Fe-rich, precipitates are observed in both cases whereas in the case of *Sus* tooth (figure 4) most of them appear to be of spherical shape filling dentinal tubules. The size of these precipitates is 5± 2.5 μm which is consistent with the size of the dentinal tubules, as they were determined from the optical microscopy images. The two larger precipitates with dimensions 10×10 μm^2 and 30×15 μm^2 located at the bottom left and top center areas of the Fe map respectively, are attributed due to their size, to precipitates filling microbial borings.

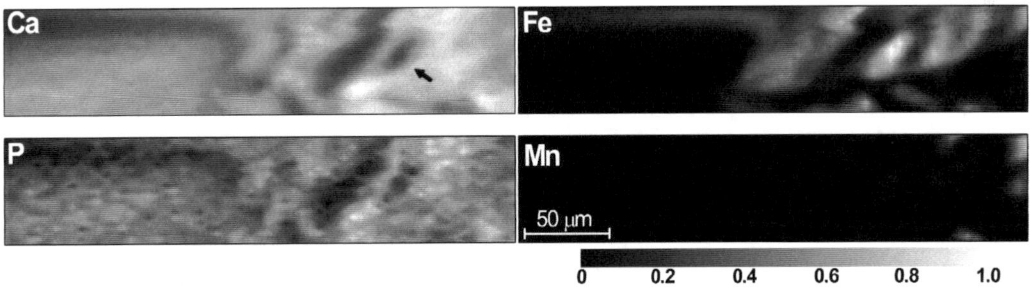

Figure 3. XRF maps of the *Hipparion* molar tooth recorded at 650 μm from the outer surface showing dentin and a region of sedimentary metal-rich precipitates. The arrow in the Ca map indicates an area of metal containing precipitates in the dentin region.

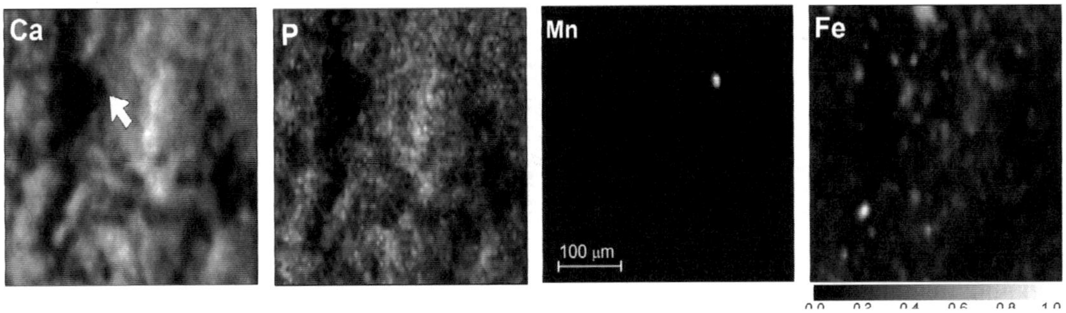

Figure 4. XRF maps of the *Sus arvernensis* canine tooth recorded at a dentin region located at 2 mm from the outer surface. The arrow in the Ca map indicates a diagenetic crack.

3.2. X-ray Absorption Fine Structure

Fe is the most abundant contaminating element detected in the cross section of all samples except for the *Ursus ingressus* tooth where the metal concentration was very low. The *Sus arvernensis* tooth pulp is surrounded by a yellowish, approximately 1.5 mm thick, layer while the remaining part has a dark brown color as observed in the cross section. In order to investigate variations in the Fe speciation, Fe *K* edge μ-XAFS spectra were recorded from Fe-rich and Fe-poor spots properly selected from the XRF maps, whereas macro-XAFS spectra were recorded from regions of different color in the samples' cross-sections. The Fe *K* edge XANES spectra shown in Figure 5 were recorded at a Fe-rich (a), Fe-poor (b), yellow-colored area (c) and brown-colored area (d) spots in the *Sus arvernensis* tooth cross-section. Fe *K* edge XANES spectra recorded at a Fe-rich spot located in the dentin region approximately 850 μm from the outer tooth surface (e), at a Fe-poor spot located in the cementum at the centre of the tooth cross-section (f) and a white region located at the lateral surface (g) of the *Hipparion* molar tooth as well as at the cross section of the *Elephas antiquus* tusk (h) are also shown.

The characteristic low intensity peak that appears before the absorption edge (hereafter it will be referred as pre-edge peak) is due to 1s→3d electronic transitions that become dipole allowed after mixing of the 3d with 4p orbitals induced by proper geometry or distortion of the Fe polyhedron [15,16]. The isolated pre-edge peaks fitted using lorentzian functions are shown in figure 5 and the fitting results are listed in table 2. During the fitting the widths of the L1 and L2 lorentzians were commonly iterated, while the width of L3 lorentzian was allowed to vary. The centroid position of the pre-edge peak, determined taking into account the positions of the L1 and L2 lorentzians that correspond to interactions of Fe with its 1st nearest neighbors, reveal that irrespectively of the local concentration of Fe or the dental region, Fe is trivalent and octahedrally coordinated [15,16]. The intensity of the L3 lorentzian corresponds to interactions of Fe involving 2nd nearest neighbors and was thus not included in the determination of the total area of the pre-edge peak and the centroid position. Slight variations in the intensity ratio of the L1-L3 components might be related to the formation of different Fe compounds and/or different degree of crystalline order. More specifically, in the *Sus* and *Hipparion* teeth we do not detect any variation in the pre-edge peak intensity between the Fe-rich and Fe-poor spots. However, the intensity of the pre-edge peak in the brown area of the *Sus* tooth and the white area of the *Hipparion* tooth is lower, revealing a higher centrosymmetry in the Fe site.

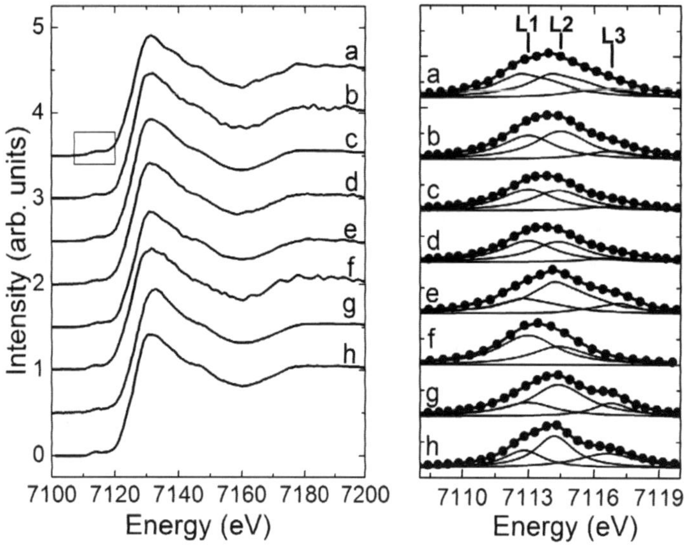

Figure 5. Fe *K* edge XANES spectra (left) and fitting of the corresponding isolated pre-edge peaks (right) of the: (a) Fe-rich spot, (b) Fe-poor spot, (c) yellow area and (d) brown area of the *Sus arvernensis* tooth; (e) Fe-rich spot, (f) Fe-poor spot and (g) white area at the lateral surface of the *Hipparion* molar tooth; (h) *Elephas antiquus* tusk. The spectra (a)–(f) and (h) were recorded at the sample cross sections and are shifted along the y-axis for clarity. The pre-edge peaks were fitted with three lorentzian components denoted as L1-L3.

Table 2. Fitting results of the pre-edge peak of the Fe *K* edge XANES spectra recorded at the cross sections of the *Sus arvernensis* canine tooth, the *Hipparion* molar tooth and the *Elephas antiquus* tusk. The Fe-rich and Fe-poor spots refer to sample positions determined by XRF mapping whereas the rest of the spectra are macro-XANES spectra recorded at characteristic sample regions. The three lorentzian functions are located at 7113.0 (L1), 7114.4 (L2) and 7116.8 (L3) eV.

Spectrum	Component	Area (arb. units)	Total area (arb. units)	Centroid position (eV)
Sus tooth (Fe-rich spot)	L1, L2, L3	0.08, 0.08, 0.03	0.16	7113.7
Sus tooth (Fe-poor spot)	L1, L2, L3	0.07, 0.08, 0.03	0.15	7113.8
Sus tooth (yellow area)	L1, L2, L3	0.07, 0.08, 0.04	0.15	7113.7
Sus tooth (brown area)	L1, L2, L3	0.06, 0.06, 0.01	0.12	7113.7
Hipparion molar tooth (Fe-rich spot)	L1, L2, L3	0.05, 0.10, 0.02	0.15	7113.9
Hipparion molar tooth (Fe-poor spot)	L1, L2	0.09, 0.06	0.15	7113.5
Hipparion molar tooth (White region at the lateral surface)	L1, L2, L3	0.04, 0.09, 0.03	0.13	7114.0
Elephas antiquus tusk random spot	L1, L2, L3	0.04, 0.07, 0.05	0.11	7113.9

The bonding geometry of Fe is further investigated by means of Fe *K* edge EXAFS spectra. The spectra were subjected to subtraction of the atomic background and transformation from the energy- to the k-space. The resulting $\chi(k)$ spectra were fitted using photoelectron scattering paths constructed using the FEFF code [12]. The $\chi(k)$ spectra and the corresponding Fourier Transforms (FT) shown in figure 6 were recorded from the yellowish and brown area and from a larger region of the *Sus* tooth cross section, from a Fe-rich spot in the dentin region near the tooth surface and from a cementum region in the *Hipparion* tooth as well as from the *Elephas antiquus* tusk. The outer surface of the *Hipparion* tooth that exhibits a white Fe-rich area was also studied. The μ-EXAFS spectrum of *Sus* tooth brown area was fitted using the photoelectron scattering paths of ferrihydrite. The μ-EXAFS spectrum of *Sus* tooth yellow area as well as the macro-EXAFS from the *Sus* pulp cavity - yellow area were fitted with the goethite model. The same model was used for the fitting of the *Elephas antiquus* tusk macro-EXAFS spectrum. The μ-EXAFS spectrum of the Fe-rich spot at the *Hipparion* tooth was fitted with the lepidocrocite model whereas the macro-EXAFS spectrum of the *Hipparion* tooth was fitted with a mixed model of 54% goethite and 46% lepidocrocite contributions. Finally, the macro - EXAFS spectrum from the white region of the *Hipparion* tooth was fitted with the vivianite ($FePO_4$) model. During the iterations the coordination numbers (N) were kept fixed according to the respective reference models. The fitting results are listed in table 3.

Iron hydroxides are common diagenetic minerals in fossils revealing a fossilization process under oxidizing environment. Goethite is the most abundant contaminating mineral detected in the samples excavated from terrestrial sites. In goethite (α-FeOOH) the FeO_6 octahedra share edges and corners with Fe-Fe distances which are approximately 3.01 (3.28) and 3.46 Å, respectively. In lepidocrocite (γ-FeOOH) the octahedra are bonded only with edges with the Fe-Fe distance equal to 3.05 Å. Ferrihydrite is a low-symmetry Fe oxide polymer where the octahedra share edges but the presence of

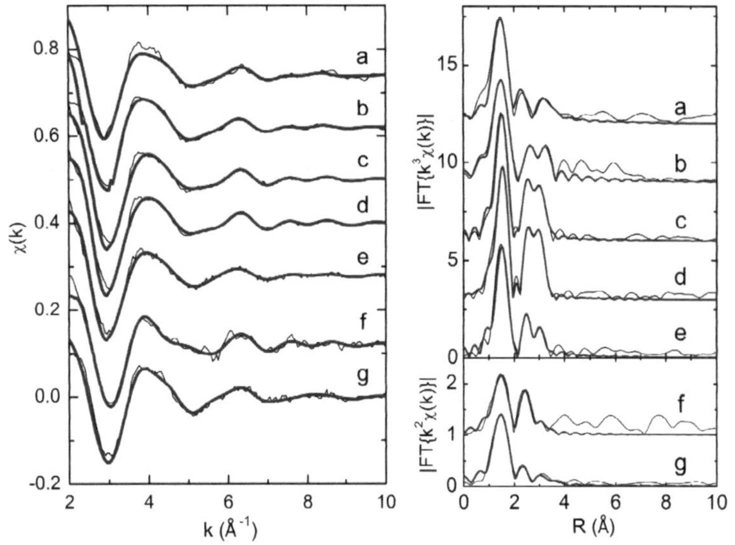

Figure 6. Fe *K* edge $\chi(k)$ (left) and corresponding amplitudes of the Fourier Transforms of the k^n weighted $\chi(k)$ spectra (right) of: (a) brown region, (b) yellow region and (c) macro EXAFS spectra of the *Sus* tooth; (d) *Elephas antiquus* tusk; (e) macro EXAFS, (f) Fe-rich spot and (g) white region of the *Hipparion* molar tooth. The experimental spectra and the fitting curves are shown in thin and thick lines, respectively.

Table 3. Fitting results of the Fe *K* edge μ- and macro-EXAFS spectra of the fossil teeth and tusk. N is the coordination number, R in Å is the nearest neighbor distance, σ^2 in $Å^2$ is the Debye-Waller factor. During the iterations the value of the coordination number N was kept fixed according to the respective reference model. The abbreviations Gt, Lp, Fh and Vn correspond to the goethite, lepidocrocite, ferrihydrite and vivianite ($FePO_4$) models.

Specimen	Parameters	Sh1	Sh2	Sh3	Sh4
Sus tooth (brown area) *Model: Fh*	N	6 O	6 Fe	-	8 O
	R	1.99±0.01	3.00±0.02	-	3.81±0.02
	$\sigma^2 \times 10^{-3}$	6.7	21.8	-	4.3
Sus tooth (yellow area) *Model: Gt*	N	6 O	4 Fe	4 Fe	10 O
	R	1.99±0.02	3.06±0.05	3.42±0.04	3.71±0.07
	$\sigma^2 \times 10^{-3}$	12.2	22.9	13.0	25.3
Sus tooth (macro-EXAFS) *Model: Gt*	N	6 O	4 Fe	4 Fe	5 O
	R	2.01±0.01	3.04±0.01	3.42±0.02	3.77±0.05
	$\sigma^2 \times 10^{-3}$	8.6	10.5	13.3	16.0
Elephas antiquus tusk *Model: Gt*	N	6 O	4 Fe	4 Fe	6 O
	R	2.01±0.01	3.04±0.01	3.41±0.01	3.80±0.04
	$\sigma^2 \times 10^{-3}$	7.83	9.41	11.07	17.5
Hipparion tooth (macro-EXAFS) *Model: 54% Gt +46% Lp*	N_{Gt}	6 O	2.15 Fe	2.15 Fe	5.37 O
	R_{Gt}	2.02±0.01	3.28±0.03	3.45±0.06	3.77±0.06
	$\sigma^2_{Gt} \times 10^{-3}$	9.9	6.4	12.2	24.7
	N_{Lp}		2.78 Fe		
	R_{Lp}		3.04±0.03		
	$\sigma^2_{Lp} \times 10^{-3}$		6.4		
Hipparion tooth (Fe-rich spot) *Model: Lp*	N	6 O	6 Fe	-	6 O
	R	2.00±0.01	3.00±0.01	-	3.74±0.06
	$\sigma^2 \times 10^{-3}$	9.5	9.1	-	8.1
Hipparion tooth (white area at the outer surface) *Model: Vn*	N	6 O	2 P	10 O	4 O
	R	1.99±0.01	3.18±0.04	3.78±0.05	4.33±0.05
	$\sigma^2 \times 10^{-3}$	4.7	8.4	18.8	2.3

voids results in a less-compact structure [17]. SEM-EDX measurements from the cementum area of the *Hipparion* molar tooth and reflected light optical images from the pulp cavity of the *Sus arvernensis* tooth revealed the presence of both pyrite and goethite in the samples. The pyrite oxidation that results in the formation of metastable minerals ferrihydrite and schwertmannite, as well as jarosite, lepidocrocite, goethite and hematite has been reported in the literature [18,19]. The detection of various polytypes of Fe oxides even in teeth fragments found in the same depositional environment could be attributed to the locally different pH values that might result in structural transitions even at room temperature [17]. More specifically, the presence of pyrite and goethite near the pulp cavity of the Milia *Sus* tooth reveals pH conditions of 4-6 during this stage of fossilization [20,21].

Contrary to the Milia and Xerias terrestrial environments, the Aridaia caves of Speleopark seem to provide an ideal fossilization environment since no considerable amounts of Fe or Mn were detected in the *Ursus* tooth cross-section. However, a black region on the external surface of the *Ursus* tooth is found to contain Mn and thus it was studied by Mn *K* edge EXAFS spectroscopy. Fitting of the $\chi(k)$ spectrum and the corresponding amplitude of the FT are shown in figure 7. The spectrum was fitted using the photoelectron scattering paths of birnessite (MnO_2) [22]. Six oxygen and six Mn atoms were found at the distance of 1.90±0.01 and 2.87±0.01 Å, respectively. The corresponding Debye-Waller factors were found equal to 2.3 and 4.6×10^{-3} Å^2, respectively. The coordination number of the first neighboring shell (Mn-O) reveals that Mn forms octahedra, while the distance of the second neighboring shell (Mn-Mn) shows that the octahedra share mainly edges. It should be pointed out that Mn oxides and hydroxides exist in various polytypes that vary in the way the MnO_6 octahedra bind together by sharing mainly edges and thus forming chains. Phyllomorph phases are formed when corner linkages do not occur while tunnel structures in 3D networks are formed by corner-sharing cross-linking of chains. The similarity of Mn bonding environment in the mid-range order of the various polytypes renders their discrimination by EXAFS difficult [23].

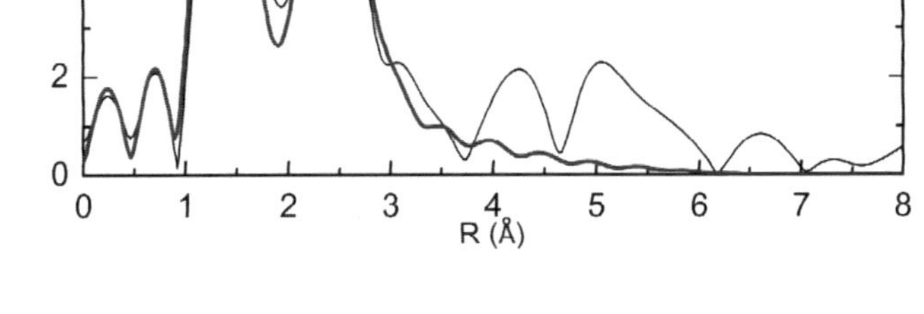

Figure 7. Fitting of the Mn *K* edge $\chi(k)$ (inset) and the corresponding amplitude of the Fourier Transform of the black region at the outer surface of the *Ursus* tooth. The experimental spectra and the fitting curves are shown in thin and thick lines, respectively.

4. Conclusions

The spatial distribution, contamination profiles and bonding environment of elements in fossil teeth and tusk excavated at sites in Northern Greece (Milia, Xerias, Almopia Speleopark) were studied by means of XAFS and XRF linear and 2D mapping. The dry sands where the Milia samples were recovered caused significant diagenetic alterations of the samples. Extensive microbial borings were detected by means of reflected light optical microscopy in the *Sus arvernensis* tooth while various Fe and Mn contaminant phases were identified in both Milia samples. The linear scans indicate that the Fe contamination paths are both through the sample surface via diffusion processes and through the empty cusps via infiltration processes. The Mn precipitations were detected only in the cementum area in the centre of the *Hipparion* tooth. The high intensity of Mn indicates the exogenous origin of Mn that–entered the sample via groundwater infiltration processes. SEM-BSE images of the *Hipparion* tooth confirm this conclusion and also reveal that the cementum area is mainly affected by metal-rich phases and to a lesser extent the dentin area around the pre- and post fossettes. The two-dimensional XRF mapping revealed the formation of Fe precipitates in the dentin region of the *Hipparion* tooth which was close to the outer surface as well as the precipitation of Fe-rich circular phases in the *Sus arvernensis* tooth dentin. Based on the size and distribution of the *Sus* Fe precipitates, as well as complementary reflected light optical microscopy images, Fe appears to fill various dentinal tubules and some microbial borings. Finally Fe *K* edge XAFS was used to assess the oxidation state as well as the coordination and bonding geometry of the metal-rich precipitates. In all the samples iron was trivalent and octahedrally coordinated. Iron hydroxides (ferrihydrite, goethite, lepidocrocite) were the only iron minerals detected in the cross-section of the samples revealing that the fossilization occurred under oxidizing conditions. Although the Xeria sample was heavily cracked along its entire length and filled with clay minerals, only iron hydroxides, and more specifically goethite, were detected. On the contrary in the Milia samples both goethite and pyrite were detected, revealing pH conditions during fossilization between 4 and 6. However the transformation was not complete. In the Almopia Speleopark sample, which was well preserved, we did not detect any considerable amounts of exogenous elements. The stable environment conditions of the cave, in addition to periodical flooding which wetted the samples and removed the organic components resulted in the complete fossilization of the sample. The burial environment affected only the external surface of the *Ursus* tooth. The structure of the black phase at the external surface of the *Ursus* tooth as well as a white phase at the external surface of the *Hipparion* tooth, were identified, without any destructive sample preparation, by means of Mn and Fe *K* edge EXAFS spectroscopy. The Pliocene lacustrine formation in which the *Hipparion* bone was found is correlated to the vivianite detected by EXAFS.

5. Acknowledgements

The measurements at the electron storage ring BESSY II of the Helmholtz Zentrum Berlin and HASYLAB-DESY were financially supported from the European Community's Seventh Framework Program (FP7/2007-2013) under grant agreement n.°226716.

6. References

[1] Ungar P S 2010 *Mammal Teeth: Origin, Evolution and Diversity* (Baltimore: The John Hopkins University Press) chapter 1 pp 9-19

[2] Lowenstam H A and Weiner S 1989 *On Biomineralization* (New York: Oxford Univ. Press) p 324

[3] Budras K D, Sack W O and Röck S 2011 *Anatomy of the horse* (Hannover: Schlütersche Verlagsgesellschaft mbH & Co KG) chapter 4 pp 32-3

[4] Eisenmann V, Alberdi M T, De Guili C and Staesche U 1988 *Studying fossil horses: collected papers after the "New York Int. Hipparion Conf., 1981"* vol 1: Methodology ed M Woodburne and P. Sondaar, (Leiden: E J Brill) pp 1–71

[5] Dauphin Y and Williams C T 2004 *C. R. Paleovol* **3** 583-90

[6] Tsoukala E and Lister A 1998 *Boll. Soc. Paleontol. Ital.* **37** 117-39

[7] Tsoukala E 2000 *Ann. Paleontol.* **86** 165–91

[8] Tsoukala E, Mol D, Pappa S, Vlachos E, van Logchem W, Vaxevanopoulos M and Reumer J 2011 *Quatern. Int.* **245** 339-49

[9] Lazaridis G 2006 *Scientific Annals, School of Geology, Aristotle University, Proc. of the 12th Int. Cave Bear Symp.* (Thessaloniki and Aridaia, GR, 2-5 November 2006) vol 98 ed E Tsoukala and G Rabeder (Thessaloniki: Aristotle University) pp 33-40

[10] Tsoukala E, Chatzopoulou K, Rabeder G, Pappa S, Nagel D and Withalm G 2006 *Scientific Annals, School of Geology, Aristotle University of Thessaloniki, Proc. of the 12th Int. Cave*

Bear Symp. (Thessaloniki and Aridaia, GR, 2-5 November 2006) vol 98 ed E Tsoukala and G Rabeder (Thessaloniki: Aristotle University of Thessaloniki) pp 41-67

[11] Bartoll J, Röhrs S, Erko A, Firsov A, Bjeoumikhov A and Langhoff N 2004 *Spectrochim Acta* B **59** 1587 -92

[12] Ankudinov A L, Ravel B, Rehr J J and Conradson S D 1998 *Phys. Rev.* B **58** 7565–76

[13] Newville M, Ravel B, Haskel D, Rehr J J, Stern E A and Yacoby Y 1995 *Physica* B **208&209** 154–6

[14] Zougrou I M, Katsikini M, Pinakidou F, Paloura E, Papadopoulou L and Tsoukala E 2014 *J Synchr. Rad* **21** 149-60

[15] Wilke M, Farges F, Petit P E, Brown Jr G E and Martin F 2001 *Am. Mineral.* **86** 714–30.

[16] Westre T E, Kennepohl P, DeWitt J G, Hedman B, Hodgson K O and Solomon E I 1997 *J. Am. Chem. Soc.* **119** 6297–314

[17] Manceau A and Drits V A 1993 *Clay Miner.* **28** 165–84

[18] Nordstrom D K 1982 *Aqueous Pyrite Oxidation and the Consequent Formation of Secondary Iron Minerals in Acid Sulfate Weathering* ed J A Kittrick, D S Fanning et al. (Madison - Soil Science Society of Amerika) chapter 3 pp 37-56

[19] Huggins F E, Huffman G P, Kosmack D. A. and Lowenhaupt D. E. 1980 *Internat. J. Coal Geol.* **1** 75-81

[20] Andreeva D, Mitov I, Tabakova T and Andreev A 1994, *J. Mater. Sci Mater.***5** 168-172

[21] Krishnamurti G S R and Huang P M, 1991 *Clay Clay Miner.* **39** 28-34

[22] Korshin G V, Chang H S, Frenker A I and Ferguson J F 2007 *Environ. Sci. Technol.* **41** 2560-5

[23] Manceau A, Gorshkov A I and Drits V A 1992 *Am. Mineral.* **77** 1133–43

Journal of Physics: Conference Series **499** (2014) 012016 doi:10.1088/1742-6596/499/1/012016

μ-SRXRF characterization of Brazilian emeralds

JF Curado[1,2], *M Radtke*[1], *G Buzanich*[1], *U Reinholz*[1], *H Riesemeier*[1], *RAS Guttler*[3] *and MA Rizzutto*[2]

[1]BAM Federal Institute for Materials Research and Testing, 12489 Berlin, Germany
[2]Institute of Physics, University of São Paulo, 05508-970, São Paulo, SP Brazil
[3]Institute of Geosciences, University of São Paulo, 05508-080, São Paulo, SP Brazil

E-mail: jcurado@usp.br, martin.radtke@bam.de

Abstract. The aim of the present study is to characterize emeralds from different mines of Brazil by using Synchrotron Radiation X-ray Fluorescence Microanalysis (μ-SRXRF). The advantage of this technique is that we can analyze a homogeneous, inclusion free area of the stone with the microbeam to distinguish the elemental fingerprint according to the provenance of the emerald. A total of 47 samples belonging to 5 different Brazilian mines were studied in this work and 28 elements were identified. By means of Principal Component Analysis (PCA) it is possible to build different groups according to the provenance of the stones, which allows to assign samples of unknown origin to the according mine.

1. Introduction

Emerald is one of the most important precious stone used in jewelry and therefore one of the most imitated gems, synthetic or simulated, as well as one of the principal items on the illegal trade of gems [1,2]. Belonging to the Beryl group, its chemical formula is Be_3Al_2 (Si_6O_{18}). Their purity, in terms of mineral inclusions, is one of the factors that determines the commercial value and also serves as proof of authenticity. The green color is associated with the presence of elements as chromium, iron, and in some cases vanadium. These trace elements can be used as fingerprint and can therefore be a guideline to determine the region of origin. Nowadays, Brazil is the largest producer, by volume, of emeralds with many important mines located mainly in the southeast and northeast of the country [3,4]. The emeralds are processed by domestic cutters or foreign buyers take the rough stones to cutting facilities in their own countries [5].

In the present study 47 natural emeralds from 5 different mines were analyzed. The occurrences are from Goiás state (Santa Terezinha de Goiás(St) and Minaçu(M)), Minas Gerais state (Nova Era(N) and Belmont(B)) and Tocantins state (Tocantins(T)).

2. Experimental Procedures

The set of stones were analyzed with μ-SRXRF (μ-Synchrotron Radiation X-Ray Fluorescence) at the BAM*line,* the beamline of the Federal Institute for Materials Research and Testing [6,7] for hard X-rays at the electron storage ring BESSYII in Berlin. The measurements were performed in homogeneous areas, free of inclusion to distinguish the elemental fingerprint of the emeralds.

The samples are small emeralds whose sizes range from 0.4 cm to 2.0 cm where some are in their raw form and others are cut having polished surface. Visually it is not possible to classify the samples according to their source. There are only few changes in the color of emeralds ranging from shades of light to dark green. The stones were all mounted on the same holder in a way that a flat surface of each stone was facing up towards the detector and on the same plane. For stones with cuts, one of the flat surfaces and for rough stones the flattest available surface was chosen for irradiation.

Excitation energies of 10 and 20 keV were used and the emitted fluorescence radiation was detected with a silicon drift detector (SDD). The acquisition time was 60s per point and the beam size was 0,5mm with a total of 4 points per sample having no relevant statistical fluctuation. Spectrum evaluation was performed with the Axil software package [8].

By using an excitation energy of 10keV we were able to identify 16 elements (Si, P, S, Cl, K, Ca, Sc, Ti, V, Cr, Mn, Fe, Ni, Cu, Zn, and Cs) and with 20 keV 12 additional elements (Ga, Ge, Se, Br, Rb, Sr, Y, Zr, Nb, W, Pb and Bi) could be detected.

Naturally, not all elements were found in each sample, some were even found in only one sample but all were present in the model used for the analysis. A typical spectrum obtained is shown in figure 1. In this figure is shown the Kα or Lα lines of some elements as example.

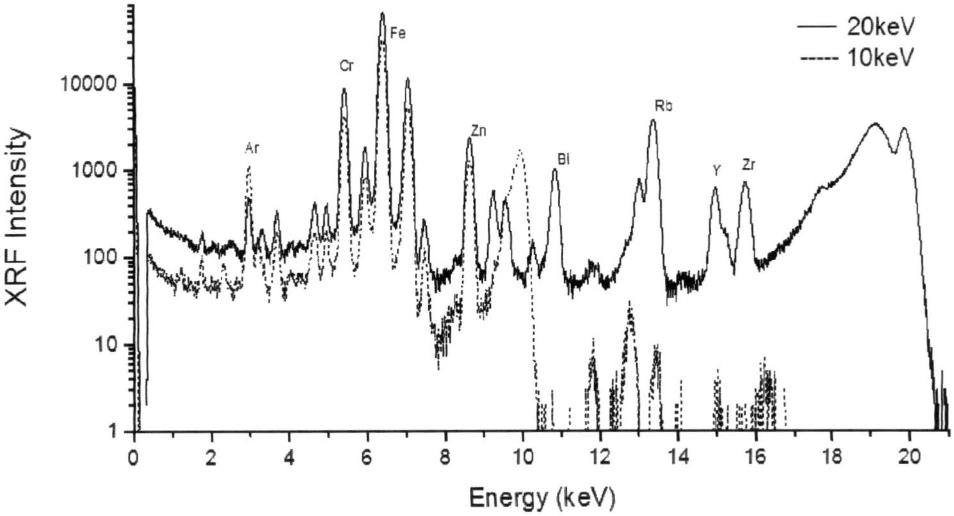

Figure 1. Typical spectra obtained by SRXRF technique for emerald sample with excitation energies of 10keV (dashed line) and 20keV (solid line).

Journal of Physics: Conference Series **499** (2014) 012016 doi:10.1088/1742-6596/499/1/012016

3. Results and Discussions

For the first analysis we generated column plots of elemental concentration for the comparison between the different samples. For some elements, such as Zn (10keV), Pb and Bi (20keV), it is possible to point to a small separation in the sample set as we can see at Figure 2. Using the results from the 10 keV measurements the group of samples originating from "Minaçu" can be separated from the other by their Zn content. With 20keV excitation energy the Bi and Pb concentrations can be evaluated. This allows the assignation of "Nova Era" and "Belmont" gems are characterized by a higher amount of Bi whereas the "Minaçu" and "Santa Terezinha" stones show higher presence of Pb in their composition.

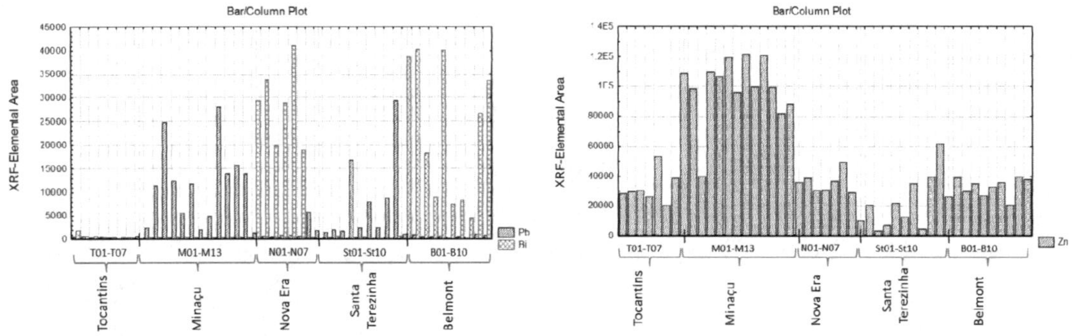

Figure 2. Column plot of intensity for emerald samples at 10 keV (Zn) and at 20 keV (Pb and Bi).

These histograms of peak areas are indicating a compositional difference between the samples from distinct places, but they are not conclusive. In order to verify the existence of the groups statistical PCA (Principal Components Analysis) was applied. The statistical analysis was carried out using the STATISTICA [9] software package. PCA reduces the number of original variables to a smaller number of representative variables, creating a factor space helping on their classification [10,11]. With these results the identification of clusters with similar characteristics with respect to these parameters is possible.

In the PCA, for the data measured with 10 keV excitation energy, the elements Ti, Mn and Zn were used as parameters. The result of this analysis is shown in the diagram in figure 3 on the left, where a separation of "Minaçu" and "Santa Terezinha" samples can be observed while there are no visible differences between the other groups. At the energy of 20 keV the elements Cr, Mn, Zn, Sr and Bi were used as parameters. In this case, a separation of the samples belonging to "Minaçu", "Santa Terezinha" and "Tocantins" can be detected, even if there are some exceptions.

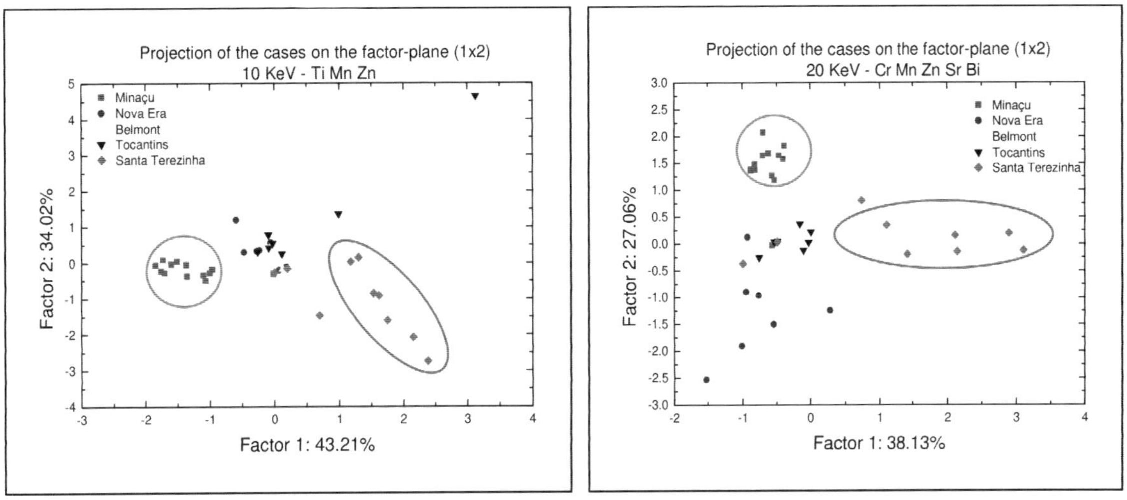

Figure 3. PCA analysis applied to SRXRF data of 57 emeralds stones. At 10keV, Ti, Mn and Zn were used as parameter. At 20 keV the elements Cr, Mn, Zn, Sr and Bi were used as parameter. The samples are listed according to their classification/mine.

The same result can be observed using a ternary graph of the elements Bi, Zn, Pb. In this graph the data was normalized by the samples to get a range between 0 and 1. The graph presents four groups which are in agreement with the information about the local origin of the stones, again with few exceptions. The samples from "Nova Era" and "Belmont" at the top of the graph form one cluster and cannot separated.

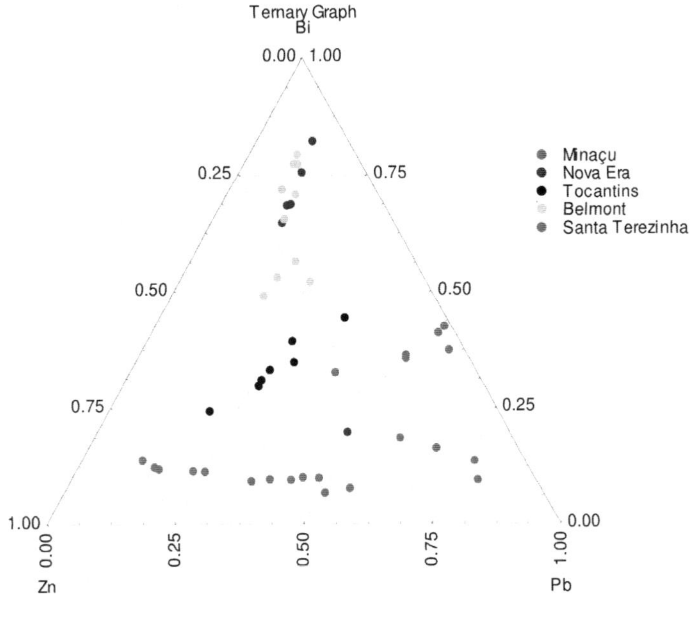

Figure 4. Ternary graph with the elements Zn, Pb and Bi at energy of 20keV.

With the data collected so for, the samples belonging to "Belmont" and "Nova Era" show a similar fingerprint. This result is conclusive given the fact that those two mines benefit from the same geological formation since they are just 10 km away from each other [12,13].

4. Conclusions

The analysis of emeralds using μ-SRXRF technique in inclusion-free areas, allows the determination of the elemental fingerprints. Characteristic elemental composition permits the assignment of each stone to the corresponding mine.

In this study, it was to differentiate the groups belonging to 3 mines, "Minaçu", "Tocantins" and "Santa Terezinha". The groups "Nova Era" and "Belmont" that are neighboring mines with the same geological formation cannot be distinguished. These results are valuable for the authentication of gems and the abatement of forgeries and the illegal trade.

Acknowledgment

The authors would like to thank the Institute of Geosciences, University of São Paulo for the providing of the samples used in this investigation, BAM and BessyII for the realization of the measurements and CNPq (Brazil) for the financial support of this work.

Reference

[1] K.Nassau, Journal of Crystal Growth 35 (1976), pg.211-222
[2] R.J.King, Geology Today, Vol.13, Issue 1, pg.37-40
[3] G Giuliani, L. J.H.D. Silva, P. Couto, Mineralium Deposita (1990), pg.57-64
[4] A.Lucas, Gems & Gemology, spring (2012), pg.73-77
[5] Puppim de Oliveira, J.A.; Ali, S.H., Resources Policy 36 (2011), pg.132.141
[6] H.Riesemeier et al, X-Ray Spectrometry, (2005) Vol34, issue2, pg.160-163
[7] W.Gorner et al, Insight, (2006) Vol48, issue9, pg 540.544
[8] P.van Espen et al, Chemometrics and Intelligent Lab. Systems, Nov (1986), pg.109-114
[9] StatSoft, Inc. (2004), STATISTICA, version 7. www.statsoft.com
[10] Jonathon Shlens, A Tutorial on Principal Component Analysis (2009)
[11] MINGOTI, S. A., Editora UFMG, 2007
[12] Preinfalk et al, J. of South America Earth Science 14 (2002) pg.867-887
[13] J. M.Watkins et al, Technical Report 56 (Relatório Técnico 56) P30_RT56

Journal of Physics: Conference Series **499** (2014) 012017 doi:10.1088/1742-6596/499/1/012017

Ageing Effects on Exhaust Gas Catalysts: Microscopic Changes Captured by X-Ray Tomography

G Hofmann[1], A Rochet[1], S Baier[1], M Casapu[1], S Ritter[2], F Wilde[3], M Ogurreck[3], F Beckmann[3] and J-D Grunwaldt[1]

[1] Institute for Chemical Technology and Polymer Chemistry, Karlsruhe Institute of Technology, Karlsruhe, Germany
[2] Institute of Structural Physics, Technical University Dresden, Dresden, Germany
[3] Institute of Materials Research, Helmholtz-Zentrum Geesthacht, Geesthacht, Germany

E-mail: grunwaldt@kit.edu

Abstract. In this work we examine the different aspects of catalyst ageing with effects ranging from the nano to the macro scale. Underlining the general importance of combining different characterisation techniques, like transmission electron microscopy (TEM), X-ray absorption spectroscopy (XAS) and X-ray diffraction (XRD) for the nanoscale, we focus on the application of X-ray absorption micro-computed tomography (micro-CT) to capture macroscopic changes in the µm to mm scale. Two series of tomographic measurements were carried out: (i) investigation of three differently treated samples by collecting one channel from a fresh, a conditioned and an aged monolith and (ii) examination of one single coated honeycomb channel with 4 wt% Pt/γ-Al_2O_3 and for comparison one with pure γ-Al_2O_3 washcoat, which have been measured in a non-destructive ex situ manner at the same position after each ageing treatment.

Main observations of the tomographic study are: (1) coating inhomogeneities between different channels taken from the same honeycomb and between different honeycombs, (2) formation of cracks in the washcoat material and (3) formation of macroscopic Pt particles in the case of 4 wt% Pt/γ-Al_2O_3 washcoat. Particularly valuable is the non-destructive ex situ investigation after different ageing steps on the same channel using X-ray tomography.

1. Introduction

Exhaust gas catalysts are an inherent part in modern combustion vehicles which remove harmful components from the exhaust gas stream [1, 2, 3]. Emission limits being progressively restricted, the ideal aftertreatment system must show very high catalytic performance, i.e. full conversion of all harmful components, and a very long operation time. In spite of major achievements, the insufficient catalytic activity at low temperatures and the durability of exhaust gas aftertreatment systems are still major issues for the automotive industry. Common mechanisms of chemical or thermal deactivation comprise sintering of active sites, fouling or poisoning. These processes require characterisation across different length scales, ranging from the exhaust gas system over the monolithic structure down to the atomic scale [3, 4].

A synthetic ageing is usually applied to simulate the effects of ageing under real conditions in a shorter time frame and being less costly [5]. Hence, we tackle here in a first step the impact of the thermal ageing for a Pt-based model monolithic diesel oxidation catalyst (DOC) [6] on different length scales. As DOC catalysts are three-dimensional objects in terms of their shape, composition, porosity and arrangement, it is reasonable to characterise them by a 3D-technique, such as X-ray tomography [7, 8].

In this paper the potential of X-ray micro-computed tomography (micro-CT) in capturing temperature induced structural and morphological changes on the micrometer scale [9, 10] is evaluated in combination with results obtained by more conventional methods such as bulk sensitive X-ray diffraction (XRD) and X-ray absorption spectroscopy (XAS) as well as transmission electron microscopy (TEM) which provide atomic scale information [11, 12]. This approach utilising X-ray absorption tomography is complementary to investigations with scanning electron microscopy (SEM), see e.g. refs. [1, 3], which beneficially allows, if combined with energy-dispersive X-ray spectroscopy (EDX), the determination of the elemental distribution in the washcoat [13]. Unfortunately, this latter technique eliminates possibilities for further studies of exactly the same sample due to its destructive sample preparation.

To demonstrate the strength of X-ray absorption tomography in this respect, two series of measurements were carried out: (i) the investigation of 4 wt% Pt/γ-Al_2O_3 coated honeycombs using three different samples by collecting one channel from a fresh, a conditioned and an aged monolith, shown in section 4.1, and (ii) using one single coated honeycomb channel from 4 wt% Pt/γ-Al_2O_3 and one from pure γ-Al_2O_3 for comparison, which have been measured at the same position after each treatment by conducting a non-destructive ex situ study as presented in section 4.2.

2. Sample Preparation, Ageing Treatments and Characterisation Methods

2.1. Sample Preparation

In a first step, cylindrical honeycomb pieces (d=20 mm, h=50 mm) were drilled out from a 400 channel per square inch cordierite monolith. The model catalyst samples, with a high surface area washcoat, were obtained by dip-coating these cylinders with 4 wt% Pt/γ-Al_2O_3, which was prepared as described in [14]. Additionally, a sample coated with pure γ-Al_2O_3 was used as reference in the ex situ study presented in section 4.2.

After drying the samples they were calcined in a static air furnace at 500 °C for 2 h, which resulted in a washcoat layer thickness of about 50 µm in average.

Figure 1. Volume rendering of a tomographic reconstruction showing a single coated honeycomb channel with dimensions $1.4 \times 1.4 \times 2.0\,mm^3$. Green: Cordierite, defining the structure of the honeycomb. Red: Washcoat, containing 4 wt% Pt/γ-Al_2O_3 as catalytic active material.

For tomographic investigations, a single channel was cut out from a honeycomb cylinder. Figure 1 shows an example for such a single channel in a tomographic reconstruction which illustrates the two regions of the catalyst coated monolith: the cordierite defining the structure (green), and the washcoat layer (red).

2.2. Ageing Treatments

The study of 3 different channels from 3 different honeycombs, series (i), includes the following samples treated in a laboratory reactor: fresh samples corresponding to samples after the above mentioned calcination step, conditioned samples (denoted "cond.") treated for 4 h at 400 °C with 4% H_2/N_2, and hydrothermally aged samples (denoted "hyd. aged") heated for 10 h at 950 °C in 10% H_2O/air.

Ex situ tomography studies, series (ii), start with a fresh honeycomb channel coated with 4 wt% Pt/γ-Al_2O_3 and a channel with pure γ-Al_2O_3, respectively. The sequence of tomography measurements and thermal treatments in a static air furnace is schematically shown in Figure 2.

Figure 2. Scheme for the sample treatment procedure during the ex situ X-ray absorption tomography study, series (ii), on a single channel: consecutive sample treatments in a static air furnace and tomography scans are performed.

This ageing procedure, from gentle influence at low temperatures to harsh conditions at highest temperatures, was chosen to benchmark the capabilities of X-ray tomography in detecting macroscopic changes in the sample. In this publication we only focus on the initial (fresh) and final state (ageing at 950 °C) to highlight the fundamental principle and to present first valuable results.

2.3. Characterisation Methods

The tomographic measurements presented in this study were performed at the "Imaging Beamline" P05 [15, 16] operated by the Helmholtz-Zentrum Geesthacht (HZG) at the synchrotron radiation source PETRA III (Deutsches Elektronen-Synchrotron, DESY, Hamburg). The undulator beamline is equipped with a Si(111) double crystal monochromator for energy filtering and a diffuser (rotating paper) to remove beam coherence. Accordingly, the sample is illuminated by a monochromatic, incoherent and parallel X-ray beam. The transmitted X-ray intensities through the sample are converted to visible light by a 300 µm thick $CdWO_4$ scintillator screen, which is magnified by an optical microscope and finally is recorded by a SciCam SC09000M CCD camera (EHD imaging). For further information on the used tomographic technique the reader is referred to a comprehensive review [17] and corresponding books [18, 19, 20] for fundamentals and principles.

X-ray absorption micro-tomography was used to measure the local absorption within the sample in three dimensions in a non-destructive way. Samples were scanned at a distance of 3-5 mm from the scintillator screen with an effective pixel size of 1.27 µm (10× magnification) and 0.2° steps over a 180° rotation at 16 keV. Reconstructions with a binning of 2 were obtained by a software maintained at P05 performing filtered back projection.

Additional analysis efforts were required for the samples measured in series (ii), since variations of sample position and orientation were encountered after remounting the samples into the sample stage after thermal treatments. This effect of "misaligned" samples is illustrated in Figure 3 (top). Therefore, before comparison and any further analysis, the corresponding data sets were "aligned" and transformed to a common 3D grid as depicted in Figure 3 (bottom). For this purpose, the built-in image registration and resampling modules (utilising Lanczos filter) from Avizo Fire 8.0 by FEI VSG3D were used [21].

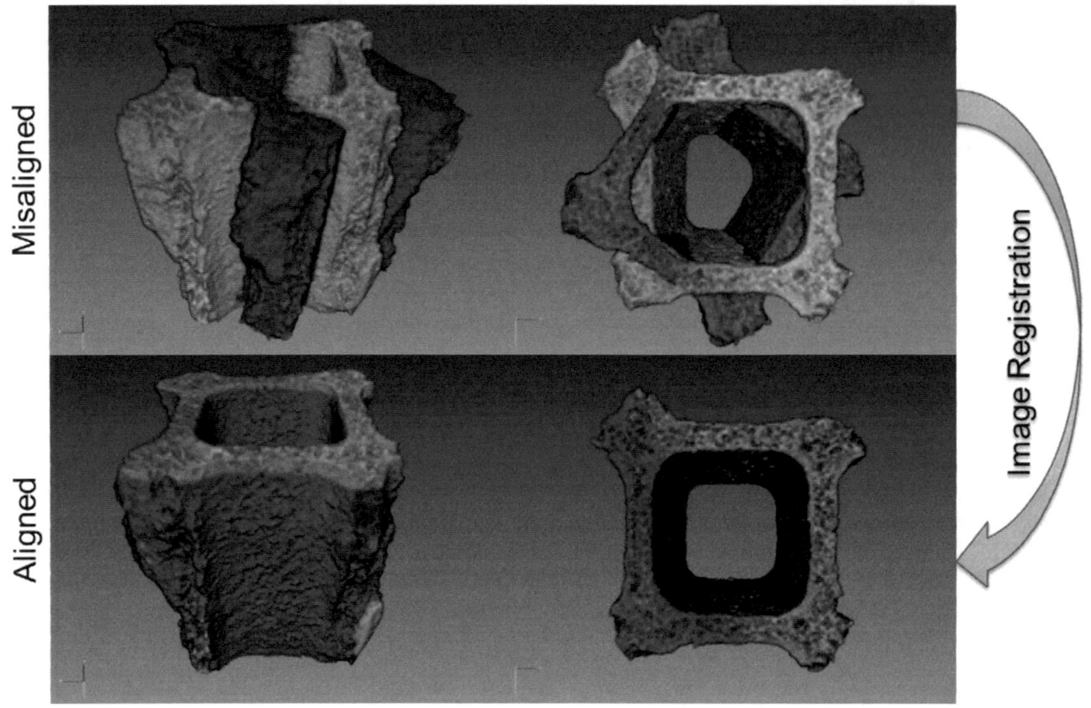

Figure 3. Consecutive sample remounting after ex situ heat treatments results in initially "misaligned" data sets shown in green and red (top). Image registration is used as the first step in data analysis to obtain "aligned" data sets (bottom).

To exclude modifications caused by sample handling, only the washcoat on walls inside a channel i.e. the red part inside the green walls in Figure 1, was considered and analysed for changes. Distinction of features between two data sets A and B is done by calculating their difference and raise it to the power of two, formally $(A - B)^2$.

For complementary analysis with more conventional techniques the sample material was collected by scratching off washcoat from fresh, conditioned or aged channels. XAS was measured on pelletised samples (diluted with cellulose) in transmission mode at the XAS beamline [22] at the ANKA synchrotron light source in Karlsruhe. The resulting spectra were energy calibrated by platinum foil as reference and normalised to pre- and post-edges using Athena from the Demeter software package [23].

TEM was carried out using a FEI Titan 80-300 aberration corrected microscope hosted at the Karlsruhe Nano Micro Facility (KNMF) at KIT. Samples were deposited on carbon coated Cu TEM grids. Data has been analysed using the Fiji software package [24].

X-ray powder diffraction patterns were obtained by using a Bruker D8 Advance X-ray diffractometer with a Cu anode operated at 35 kV and 45 mA. Data was background subtracted using the EVA software provided by Bruker and stacked by 1000 counts but kept unchanged in scaling for plotting.

Journal of Physics: Conference Series **499** (2014) 012017 doi:10.1088/1742-6596/499/1/012017

3. Characterisation Results for Probing the Structure on a Nanoscale

The decrease of the catalytic activity during CO-oxidation and related reactions in exhaust gas catalysis due to thermal ageing is well known and has been previously reported in literature, e.g. [3, 14, 25]. For the Pt model catalyst used in this study the results of the XAS, TEM and XRD measurements are summarised in Figure 4.

The intensity and profile of the whiteline feature at the Pt L_3 edge (Figure 4a) reveal the presence of platinum oxide in the fresh catalyst, partial reduction of the conditioned sample and formation of metallic Pt after ageing (spectra superimposable to Pt foil reference spectra). The analysis of the EXAFS region (not shown) confirmed the formation of very large Pt particles after sintering at 950 °C.

This is in line with the TEM data (Figure 4b) which uncovered the presence of very small particles in the fresh catalyst (80% around 1.5 nm) and particle sizes of mainly 2 nm after conditioning. The particle size distribution changes significantly for the 950 °C aged catalyst, where only 40% of the particles remain below 2 nm and much larger sizes from 12 to 350 nm were found for 7% of the particles (not shown).

Using XRD we found some cordierite due to the above mentioned sample preparation. Together with Al_2O_3 it appears additionally to the labelled Pt pattern in the diffractograms (Figure 4c). The characteristic reflections stemming from metallic platinum are exclusively visible in the hydrothermally aged sample. This confirms the results obtained from TEM and XAS i.e. the formation of large and metallic Pt particles. All three characterisation techniques confirm the occurrence of a sintering process during ageing at high temperature that is observable on the nanoscale. However, changes in the system on the micrometer scale remain hidden to the above applied methods.

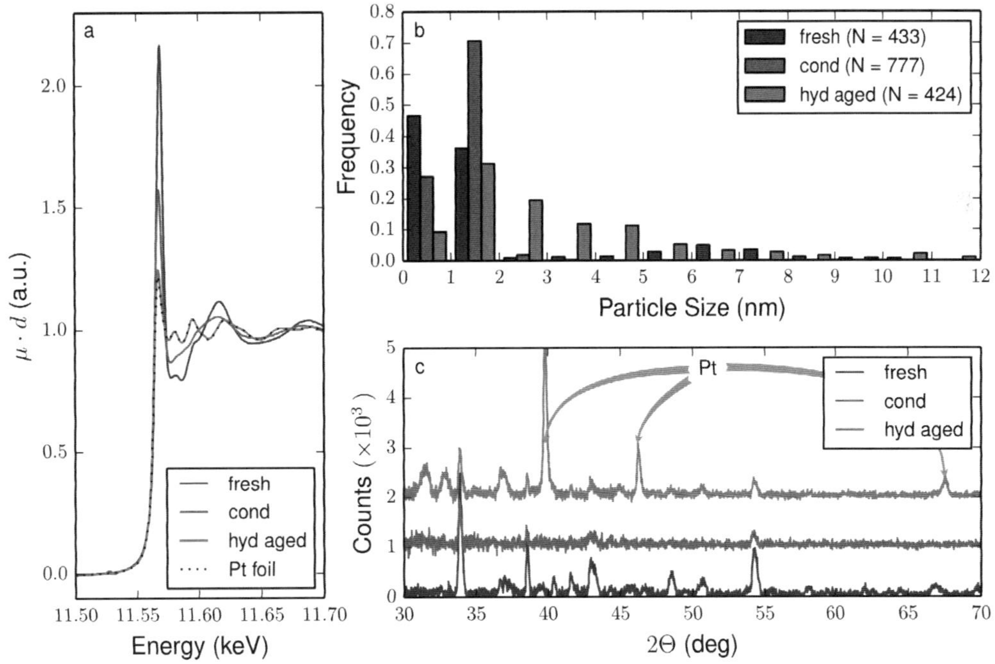

Figure 4. Characterisation results for the fresh, conditioned, and hydrothermally aged samples. **a)** XANES spectra showing the Pt L_3 edge of the samples and Pt foil as reference. **b)** Resulting particle size distribution from TEM by analysing a number of N Pt particles. **c)** Corresponding X-ray diffractograms.

4. Tomography Results

4.1. Monitoring of 3 Different Laboratory Prepared Channels

For the first series of measurements, series (i), three different channels were cut out from the corresponding fresh, conditioned and hydrothermally aged monoliths containing a 4 wt% Pt/γ-Al_2O_3 washcoat. Figure 5 presents for each sample a tomographically reconstructed slice perpendicular to the longitudinal channel axis. It can be observed that the cordierite has a very high porosity, which is maintained for all treatment steps, since it has been annealed at very high temperatures during fabrication.

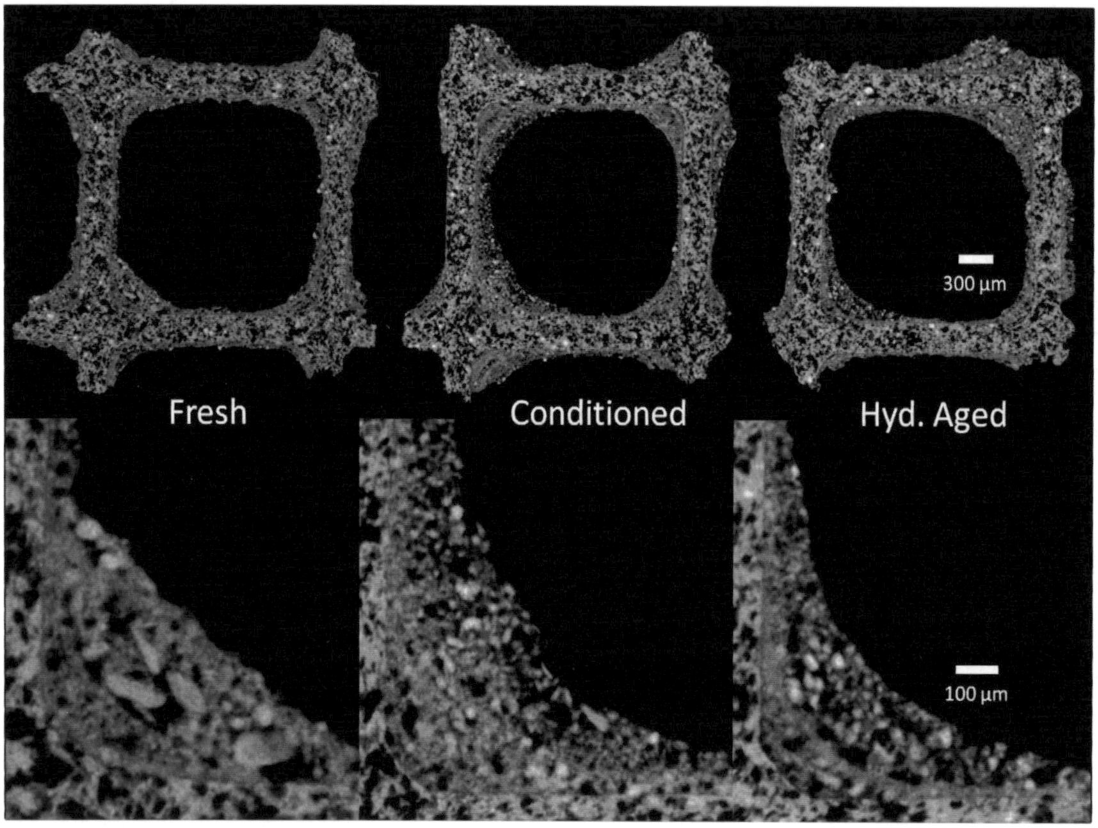

Figure 5. Three different honeycomb channels in fresh, conditioned and hydrothermally aged state. Top row: virtual cut perpendicular to the longitudinal channel axis showing the complete $1.4 \times 1.4\,\mathrm{mm}^2$ channel. Bottom row: magnification of the lower left corner. Green: Cordierite, defining the structure of the honeycomb. Red: Washcoat, containing 4 wt% Pt/γ-Al_2O_3. Black: void. Higher brightness corresponds to higher absorption.

Concerning the washcoat depicted in Figure 5, variation of the coating thickness and distribution is observed within and among the channels. This mainly originates from the dip-coating process during sample preparation and illustrates the limitation in obtaining a very homogeneous catalyst layer with the rather simple preparation method used here. This has effects on the utilisation of the catalyst material in the sample [26] and especially complicates comparison between samples. Moreover, the presence of smaller and brighter grains was observed, especially in the hydrothermally aged sample. As higher brightness corresponds to higher X-ray absorption, this could potentially be caused by sintering of the alumina carrier or, more likely, by Pt particles. Nevertheless, since different channels are scanned, it is difficult to draw a clear conclusion on the origin of the mentioned features, except if a large number of samples would be investigated to obtain sufficient statistics. This emphasizes the need to study single channels in sequence at the same position as presented in the next section.

4.2. Ex situ Tomography Monitoring a Single Channel

To capture effects of thermal treatments on the sample, we need to investigate exactly the same sample before and after each treatment as evident from the above study. Therefore we use one single coated honeycomb channel with pure γ-Al_2O_3 and one with 4 wt% Pt/γ-Al_2O_3 washcoat. Thermal treatments and measurements of the samples are performed as schematically shown in Figure 2. On the obtained data we applied the above mentioned image registration as illustrated in Figure 3.

Here we compare the state of the fresh and the 4 h at 950 °C aged sample (cf. experimental part) for both mentioned samples. To exclude accidental modifications, as mentioned above, we only analyse the washcoat inside the channel. Therefore we apply a virtual clipping to the data as shown in Figure 6, to obtain relevant parts of the sample and a free view on a single wall.

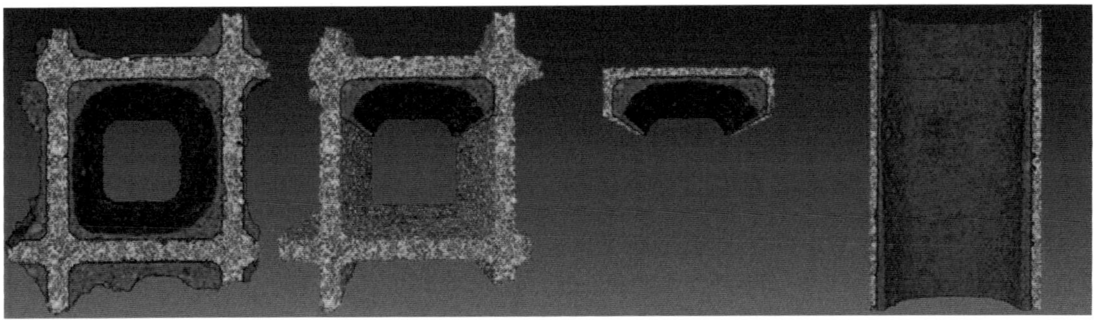

Figure 6. Illustration of virtual data clipping as used in the following discussion. To exclude accidental sample modifications, only the washcoat (red) inside the channel (green) is considered during evaluation. We apply virtual clipping to obtain a free view on the region of interest.

4.2.1. Ageing of γ-Al_2O_3 coated Channel - Evidence for Crack Formation

For the pure γ-Al_2O_3 coated reference sample after ageing treatments, we observed very strong formation of cracks and changes in the coating morphology along the channel, which originates from material displacement (Figure 7).

Crack formation already starts at temperatures as low as 600 °C and increases as a function of temperature. Thus, most distinct effects are obtained for ageing at 950 °C, as shown in Figure 7. Most of the differences (marked in red in Figure 7) between the fresh and aged samples are positioned in the vicinity of the cracks and no indication of distinct particle formation (higher density) on the plane area was found. These first observations underline that the morphological changes within the washcoat can be followed by X-ray absorption micro-tomography. This demonstrates the potential of the technique vividly.

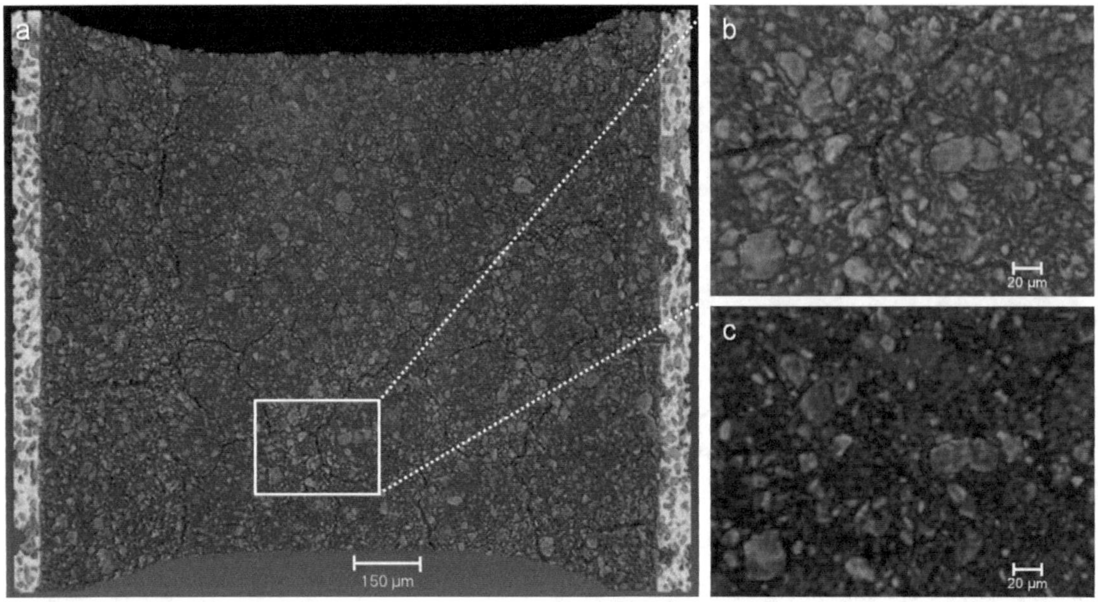

Figure 7. Direct overlay of initial (fresh, green) and final (ageing at 950 °C, gray) state for the pure γ-Al_2O_3 coated sample, where large differences are coloured in red. **a)** View on a wall viewed from inside the channel by applying a virtual clipping to the volume rendering parallel to the wall. Field of view $1.4 \times 1.1\,mm^2$. **b)** Zoom to region with formation of cracks. The extent and direction of material displacement due to the formation of the cracks can be followed by grains partially rendered in green (initial) and white (final) indicating the movement. **c)** For comparison the same area of the fresh sample without any crack.

4.2.2. Ageing of 4 wt% Pt/γ-Al_2O_3 coated Channel - Evidence for Particle Formation

In contrast to the pure alumina washcoat, the Pt containing sample shows crack formation to a much lesser extent (Figure 8). The location of the cracks is mostly at the corners of the channel where the layer thickness is biggest, as observed in Figure 5. In addition, numerous pronounced differences (marked in red in Figure 8) emerge due to the formation of larger particles. These regions with significantly higher amount of Pt are reliably detected by increased absorption as compared to the fresh sample. The results in Figure 8 show that their evolution is clearly not related to the crack formation as observed for the γ-Al_2O_3 channel. However, particle formation is not homogeneous across the sample, which may be due to variations in the Pt dispersion and distribution within the washcoat. The Pt-rich particles are already detected at 750 °C (not shown here) and grow further in size and number as temperature increases up to 950 °C in accordance to literature. These results are in line with the sintering behaviour uncovered by the XAS, TEM and XRD data but moreover these variations observed by micro X-ray absorption tomography show that thermal deterioration can also be observed on a macroscale, predicting a strong decrease in catalyst efficiency.

4.3. Comparison of results from series (i) and series (ii)

Comparing series (i) and series (ii) we first need to emphasize the methodical difference in conducting the experiments. In series (i) several different samples with diverse treatments were investigated. Differences in the washcoat amount and position can be seen immediately, as well as grains forming the washcoat and indications of sintered Pt particles. Statements on washcoat grains, their origin or evolution can hardly be made and there is a risk of misinterpretation,

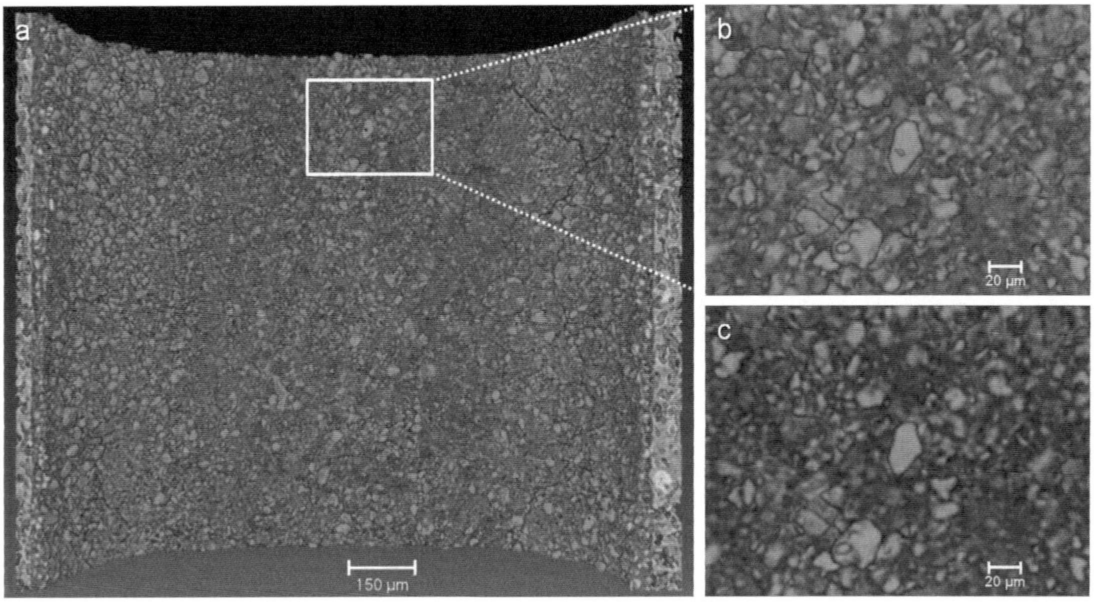

Figure 8. Direct overlay of initial (fresh, green) and final (ageing at 950 °C, gray) state for the Pt containing sample, where large differences are coloured in red. **a)** View on a wall viewed from inside the channel by applying a virtual clipping to the volume rendering parallel to the wall. Field of view $1.4 \times 1.1\,\mathrm{mm}^2$. **b)** Zoom to a region with particle formation. The position of the micrometer sized grains of the initial substrate remains unchanged in contrast to the uncoated sample. Newly formed particles are marked in red. **c)** For comparison the same area of the fresh sample.

since the initial state is unknown. Nevertheless it can be useful if these properties, like amount and distribution, have to be investigated.

Experiments in series (ii) are performed on exactly the same channel, which is subject to consecutive thermal treatments, to address the shortcoming of series (i). Consequently it allows to deduce statements of feature origin and evolution in the washcoat. On the other hand, no comparison to multiple other channels is done and has to be considered.

Both series provide equivalent information on the washcoat amount and distribution. In addition, series (ii) allows to reliably pinpoint sintered Pt particles and follow material movements as a function of the applied treatment.

The present study shows that such studies should be promoted in future where micro-tomography in X-ray absorption [7, 27], fluorescence [28, 29] or diffraction [30, 31] contrast is used on the same sample after different treatments or preferentially even in situ and thereby in short time intervals as reported in ref. [30].

5. Conclusions

This study of a model exhaust gas catalyst containing a 4 wt% Pt/γ-Al_2O_3 washcoat subjected to thermal ageing has demonstrated ageing effects both on the nano- and macroscale leading to catalyst deactivation. We have confirmed by XAS, TEM and XRD that, in accordance with literature, ageing alters properties on the nanometer scale like increasing the noble metal particle size distribution. Effects on the macroscale have been captured by non-destructive X-ray absorption micro-tomography of single honeycomb channels. Here monitoring of the same channel is of paramount importance to draw reliable conclusions.

Finally our present study revealed:

(i) Large variation of the amount and distribution of deposited washcoat on the channels, which is attributed to sample preparation. This affects the accessibility of catalyst material in the sample, controlled by mass transport, and especially complicates comparison between samples.

(ii) The formation of cracks in the washcoat material, leading to decrease in efficiency and in the worst case to rapid loss of material in the channel.

(iii) Particle sintering and appearance of micrometer sized particles in the 4 wt% Pt/γ-Al_2O_3 washcoat at 950 °C. This implies a rapid decrease in the number of small particles and thus potentially loss of highly active catalytic sites.

(iv) Spatially non-uniform formation of particles across the sample, which may be due to variations in the Pt dispersion and distribution within the washcoat.

These findings can be used in general to improve catalyst design and preparation. By applying advanced preparation methods, one can achieve more uniform and reproducible coating results for optimised catalyst utilisation. Enhanced stability and durability can be gained by introducing supports which are less prone to sintering and crack formation in conjunction with refined platinum dispersion in the coating suspension. Moreover, tomography can be used for characterisation of multi-layer catalysts to improve performance and to give input and feedback to the "product design" process to obtain long lasting and highly active catalysts.

In situ X-ray tomography with such samples under reactive conditions (gas flow, temperature) is highly desirable and is in principle possible due to the high penetrating nature of hard X-rays although dedicated in situ reactors still have to be developed. Besides the determination of the local absorption, different characteristics of the sample can be addressed by choosing appropriate tomographic contrasts. These are for example fluorescence for mapping the elemental distribution, diffraction for distinguishing crystalline phases or phase-contrast for investigating weakly absorbing samples. Altogether this renders the technique very versatile and capable of providing complementary information to other characterisation techniques, e.g. electron microscopy. Especially the combination of electron microscopy and X-ray microscopy seems to be an excellent tool despite some efforts must still be made to bridge the resolution gap for in situ tomographic imaging using hard X-rays with a spatial resolution of 20 to 500 nm.

Acknowledgements

Georg Hofmann appreciates funding by the Helmholtz Research School "Energy-Related Catalysis". Moreover, we thank the BMBF-project "X-ray microscopy" (05K10VK1) and "Nanoscopy" (05K13VK2) and the virtual institute VI-403 for financial support. Furthermore, we are grateful for beamtime at ANKA and PETRA III. Finally, we thank Elen Ogel for support of the synchrotron measurements and Di Wang for TEM measurements.

References

[1] Votsmeier M, Kreuzer T, Gieshoff J and Lepperhoff G 2000 *Automobile Exhaust Control. Ullmann's Encyclopedia of Industrial Chemistry* (Wiley-VCH) pp 407–424 URL `http://dx.doi.org/10.1002/14356007.a03_189.pub2`

[2] Heck R M, Farrauto R J and Gulati S T 2009 *Catalytic air pollution control: Commercial technology* 3rd ed (Hoboken and N.J: John Wiley) ISBN 0470275030

[3] Deutschmann O and Grunwaldt J-D 2013 *Chemie Ingenieur Technik* **85** 595–617 URL `http://dx.doi.org/10.1002/cite.201200188`

[4] Bell A T 2003 *Science* **299** 1688–1691 URL `http://dx.doi.org/10.1126/science.1083671`

[5] Andersson J, Antonsson M, Eurenius L, Olsson E and Skoglundh M 2007 *Applied Catalysis B: Environmental* **72** 71–81 ISSN 09263373 URL `http://dx.doi.org/10.1016/j.apcatb.2006.10.011`

[6] Russell A and Epling W S 2011 *Catalysis Reviews* **53** 337–423 URL http://dx.doi.org/10.1080/01614940.2011.596429
[7] Grunwaldt J-D and Schroer C G 2010 *Chemical Society Reviews* **39** 4741–4753 URL http://dx.doi.org/10.1039/c0cs00036a
[8] Grunwaldt J-D, Wagner J B and Dunin-Borkowski R E 2013 *ChemCatChem* **5** 62–80 URL http://dx.doi.org/10.1002/cctc.201200356
[9] Gibson E K, Zandbergen M W, Jacques, Simon D M, Biao C, Cernik R J, O'Brien M G, Di Michiel M, Weckhuysen B M and Beale A M 2013 *ACS Catalysis* **3** 339–347 URL http://dx.doi.org/10.1021/cs300746a
[10] Beale A M, Jacques, Simon D M and Weckhuysen B M 2010 *Chemical Society Reviews* **39** 4656–4672 URL http://dx.doi.org/10.1039/C0CS00089B
[11] Niemantsverdriet J W 2007 *Spectroscopy in catalysis: An introduction* 3rd ed (Weinheim: Wiley-VCH) ISBN 9783527316519
[12] Grunwaldt J-D, Kimmerle B, Baiker A, Boye P, Schroer C G, Glatzel P, Borca C N and Beckmann F 2009 *Catalysis Today* **145** 267–278 URL http://dx.doi.org/10.1016/j.cattod.2008.11.002
[13] Winkler A, Ferri D and Aguirre M 2009 *Applied Catalysis B: Environmental* **93** 177–184 URL http://dx.doi.org/10.1016/j.apcatb.2009.09.027
[14] Boubnov A, Gänzler A, Conrad S, Casapu M and Grunwaldt J-D 2013 *Topics in Catalysis* **56** 333–338 URL http://dx.doi.org/10.1007/s11244-013-9976-6
[15] Haibel A, Beckmann F, Dose T, Herzen J, Ogurreck M, Müller M and Schreyer A 2010 *Powder Diffraction* **25** 161–164 URL http://dx.doi.org/10.1154/1.3428364
[16] Haibel A, Ogurreck M, Beckmann F, Dose T, Wilde F, Herzen J, Müller M, Schreyer A, Nazmov V, Simon M, Last A and Mohr J 2010 *Proc. of SPIE* (*Developments in X-Ray Tomography VII* vol 7804) p 78040B URL http://dx.doi.org/10.1117/12.860852
[17] Landis E N and Keane D T 2010 *Materials Characterization* **61** 1305–1316 URL http://dx.doi.org/10.1016/j.matchar.2010.09.012
[18] Avinash C Kak, Malcolm Slaney 2001 *Principles of Computerized Tomographic Imaging* (IEEE Press) ISBN 9780898714944 URL http://www.slaney.org/pct/
[19] Buzug T M 2008 *Computed tomography: From Photon Statistics to Modern Cone-Beam CT* (Berlin and Heidelberg: Springer) URL http://dx.doi.org/10.1007/978-3-540-39408-2
[20] Herman G T 2009 *Fundamentals of Computerized Tomography: Image Reconstruction from Projections* Advances in Pattern Recognition. (Springer London) URL http://dx.doi.org/10.1007/978-1-84628-723-7
[21] Avizo Fire 8 - 3D Analysis Software for Materials Science URL http://www.vsg3d.com/avizo/fire
[22] Grunwaldt J-D, Hannemann S, Göttlicher J, Mangold S, Denecke M A and Baiker A 2005 *Physica Scripta* **2005** 769 URL http://dx.doi.org/10.1238/Physica.Topical.115a00769
[23] Ravel B and Newville M 2005 *Journal of Synchrotron Radiation* **12** 537–541 URL http://dx.doi.org/10.1107/S0909049505012719
[24] Schindelin J, Arganda-Carreras I, Frise E, Kaynig V, Longair M, Pietzsch T, Preibisch S, Rueden C, Saalfeld S, Schmid B, Tinevez J Y, White D J, Hartenstein V, Eliceiri K, Tomancak P and Cardona A 2012 *Nature Methods* **9** 676–682 URL http://dx.doi.org/10.1038/nmeth.2019
[25] Wiebenga M H, Kim C H, Schmieg S J, Oh S H, Brown D B, Kim D H, Lee J H and Peden C H 2012 *Catalysis Today* **184** 197–204 URL http://dx.doi.org/10.1016/j.cattod.2011.11.014
[26] Roy S, K Heibel A, Liu W and Boger T 2004 *Chemical Engineering Science* **59** 957–966 ISSN 00092509 URL http://dx.doi.org/10.1016/j.ces.2003.12.001
[27] Schroer C G, Kuhlmann M, Gunzler T F, Lengeler B, Richwin M, Griesebock B, Lützenkirchen-Hecht D, Frahm R, Ziegler E, Mashayekhi A, Haeffner D R, Grunwaldt J-D and Baiker A 2003 *Applied Physics Letters* **82** 3360–3362 URL http://dx.doi.org/10.1063/1.1573352
[28] Samber B, Silversmit G, Evens R, Schamphelaere K, Janssen C, Masschaele B, Hoorebeke L, Balcaen L, Vanhaecke F, Falkenberg G and Vincze L 2008 *Analytical and Bioanalytical Chemistry* **390** 267–271 URL http://dx.doi.org/10.1007/s00216-007-1694-0
[29] Basile F, Benito P, Bugani S, Nolf W d, Fornasari G, Janssens K, Morselli L, Scavetta E, Tonelli D and Vaccari A 2010 *Advanced Functional Materials* **20** 4117–4126 URL http://dx.doi.org/10.1002/adfm.201001004
[30] Jacques, Simon D M, Di Michiel M, Beale A M, Sochi T, O'Brien M G, Espinosa-Alonso L, Weckhuysen B M and Barnes P 2011 *Angewandte Chemie International Edition* **50** 10148–10152 URL http://dx.doi.org/10.1002/anie.201104604
[31] Álvarez-Murga M, Bleuet P and Hodeau J L 2012 *Journal of Applied Crystallography* **45** 1109–1124 URL http://dx.doi.org/10.1107/S0021889812041039

Journal of Physics: Conference Series **499** (2014) 012018 doi:10.1088/1742-6596/499/1/012018

Phase Contrast X-Ray Synchrotron Microtomography for Virtual Dissection of the Head of *Rhodnius prolixus*

G Sena[1], A P Almeida[1], D Braz[1], L P Nogueira[2], M V Colaço[2], J Soares[2], S C Cardoso[3], E S Garcia[4], P Azambuja[4], M S. Gonzalez[5], S Mohammadi[6,7], G Tromba[7] and R C Barroso[2]

[1] COPPE/Federal University of Rio de Janeiro, Brazil
[2] Medical Physics Laboratory/Physics Institute/State University of Rio de Janeiro, Brazil
[3] Physics Institute/Federal University of Rio de Janeiro, Brazil
[4]Laboratory of Biochemistry and Physiology of Insects/Oswaldo Cruz Institute/ FIOCRUZ, Brazil
[5] Department of General Biology/Federal University Fluminense, Brazil
[6]Department of Physics, University of Trieste, Italy
[7]Elettra - Sincrotrone Trieste ScpA, Trieste, Italy

E-mail: apalmeid@gmail.com

Abstract. Phase Contrast X-Ray Synchroton Microtomography is a non-destructive technique that allows the microanatomical investigations of *Rhodnius prolixus*, one of the most important insect vectors of Trypanosoma cruzi. In this work complete series of virtual thin sections through the heads of selected *Rhodnius prolixus* were obtained. The sections of the head were important to compare the difference in use the spatial resolution of 2μm or 4.5 μm and to see anatomical details that couldn´t be seen with other technique. Three different groups of *Rhodnius prolixus* were used. One group was fed with defibrinated rabbit blood and after 10 days was sacrificed, other group was sacrificed 4 days after feeding and the last group remained unfed. The results show some differences for each kind of groups and for the different resolutions.

1. Introduction

Conventional X-ray computed tomography (CT) is based on the difference in radiation absorption by different materials or tissues. At the same time, a wide range of samples used in materials science, biology, and medicine demonstrate very weak absorption contrast, nevertheless producing significant phase shifts in the X-ray beam. The use of phase information for imaging purposes is therefore a suitable alternative.

Application of X-ray microtomography (μCT) on insects is quite recent [1, 2, 3, 4, 5] and its transposition to use phase-contrast synchrotron X-ray microtomography is even more recent [6, 7, 8, 9]. Phase contrast X-ray imaging is already shown to be useful to study other insects within different matters.

Members of the subfamily *Triatominae*, family *Reduviidae*, are an interesting group to study for comprising a large number of insect species of which some are vectors of *Trypanosoma cruzi*, the causative agent of Chagas disease. *Rhodnius prolixus* is a blood-sucking insect well known by for its importance as vector of Chagas disease, especially in Latin America. Among the parasitic diseases, Chagas disease is ranked as one of the most important in Latin America in terms of social and economic impact, affecting about 18 million people, with about 100 million people living in what are considered to be high risk zones, and approximately 300,000 new cases occurring every year with around 21,000 deaths annually[10].

Despite the fact that the *R. prolixus* is one of the most well-know model in terms of both physiology and vector-parasite interactions studies, the understanding of how some morphophysiological aspects of mainly the retrocerebral complex in different time of the life cycle are related to the development and reproduction of this insect vector remains remarkably limited. Head morphology is also largely unknown, especially internal features. It was demonstrated in earlier studies that head structures are not only of great importance in terms of coordination of the central nervous system, sensory perception, and food processing, but also provide much phylogenetic information due to their complexity [3].

The demand for this kind of investigation is increasing with the necessity to find new vector control methods. The present study aimed to visualize three-dimensional structures at micrometer scale of the head of *R. prolixus* defining their precise spatial relationships, variations in size, shape and proportion of individual structures.

This work is part of a series of articles that use Synchrotron Radiation Phase Contrast microtomography (SR-PhC-μCT) for the study of *R. prolixus* morphology. In the first paper [11] the visualization of various detailed features were highlighted thanks to the edge enhancement effects typical of the phase contrast technique, but the contrast between foreground and background remained poor. In the second study [12] the same data set have been revisited with application of a single distance phase retrieval algorithm for comparison with previous results. The resulting slices showed very high quality images that enable a better visualization of important muscles and neurohemal organs of the central nervous system within the head of *R. prolixus*.

In the present work, the μCT images were obtained using the new experimental set-up which was recently made available at the SYRMEP beamline [13] and these images were compared to those used in the first and second papers. The new set-up uses the polychromatic configuration. In this configuration the outcoming beam from the ring is intercepted before the monochromator and in a dedicated end-station, absorption and phase contrast radiography and tomography set-up are available. A 2 μm resolution was achieved in the μCT images with the set-up used for these analyses. The images obtained with the new set-up at SYRMEP beamline provided a better visualization of virtual sections of the head of *R. prolixus*.

2. Materials and methods

2.1. Sample preparation

Fifth-instar nymphs of R. prolixus (n=5) were taken from colony in the Laboratory of Biochemistry and Physiology of Insects, Oswaldo Cruz Foundation, Brazil, maintained at 28°C and between 60% and 70% relative humidity. The insects were fed with rabbit blood using a membrane apparatus previously described by Garcia et al. [14]. Three days after blood meal, insects were immobilized at 4°C for 10 min and bounded on a polystyrene table with entomological pins and transversally cut at the junction between prothorax and mesothorax segments of body. The anterior fragments were fixed (gluteraldehyde) and maintained at room temperature.

For the X-ray microtomography (μCT) investigation the head and prothorax were dried to avoid shrinking artifacts caused by water loss during the tomography procedure, which might affect fresh samples exposed to the dry atmosphere in the experimental hutch. This procedure is usual in studies with insects[2, 3, 5, 9]. All specimens (n = 5) were imaged.

2.2. Experimental set-up at SYRMEP beamline

The first microCT images of the entire internal anatomy of the head of specimens of R. prolixus were obtained using the microCT setup at the SYRMEP (Synchrotron Radiation on Medical Physics) beamline of the ELETTRA synchrotron radiation facility (Trieste, Italy). The experimental station is placed about 23m and at this distance, the beamline provides a monochromatic laminar-section X-ray with a maximum area of about 160x5 mm2 at 20 keV. The detector system is comprised of a 12/16-bit CCD camera, with 4008 x 2672 pixels2, 4.5 µm pixel size with a field of view of 18x12 mm2, coupled to an intensifier screen with no magnification (1:1).

To gather as much information as possible from the specimens of R. prolixus, we decided to use the new high-resolution microCT setup which has been recently available at the SYRMEP for this study. In order to optimize the performances of the microCT setup for high resolution experiments in the SYRMEP beamline, a lens-coupled CCD camera system (high-resolution 2048 x 2048 pixels2, effective pixel size of 1.03 µm) designed to achieve up to 2 µm spatial resolution was used in white/pink X-ray beam mode that provides a nearly parallel, laminar-section X-ray beam with a maximum area of 100x6 mm2 at a distance of about 15 m from the source. Although monochromatic beam was not possible to be used in this high-resolution setup at SYRMEP beamline, physics filters of Si (1 mm) and Mo (0.025 mm) were used to minimize beam-hardening. Another advantage of using this new setup was the faster acquisition time achieved. For our experiments, 1440 radiographic projections were acquired over an angular range of 180° with angular step of 0.125°.

Tomographic slices were reconstructed using the conventional filtered backprojection algorithm with Shepp Logan filter using the SYRMEP TOMO PROJECT (STP) software developed by the SYRMEP team. ImageJ® software was used to render the 3D volumes.

3. Results and discussions

Morphological studies on insects often focus on a detailed analysis of the head or parts of it. The insect head shows a broad spectrum of structures, such as sensory organs, endo and exo skeletal features, muscles, brain or mouthparts, whose specific organization can act as an aid to understanding the evolution, development and ecology of the insect under study [15, 16, 17]. In this work, we obtained complete series of virtual thin sections through the heads of selected Rhodnius prolixus; these could be used both for the inspection of planar sections and for 3D reconstructions of the entire head. In figure 1, it can be seen anatomical details of the compound eyes such as omatidia and the axial filament, which could not be revealed from our previous images [11, 12].

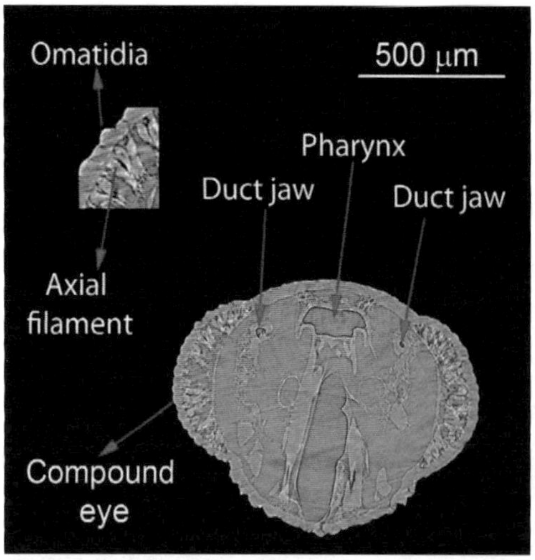

Figure 1. Virtual section (transversal view) through the head of *R. prolixus*.

The pharynx and its dilator muscles were recognized in figures 2 to 4. Variation in size of pharynx can be noted in different times after feeding.

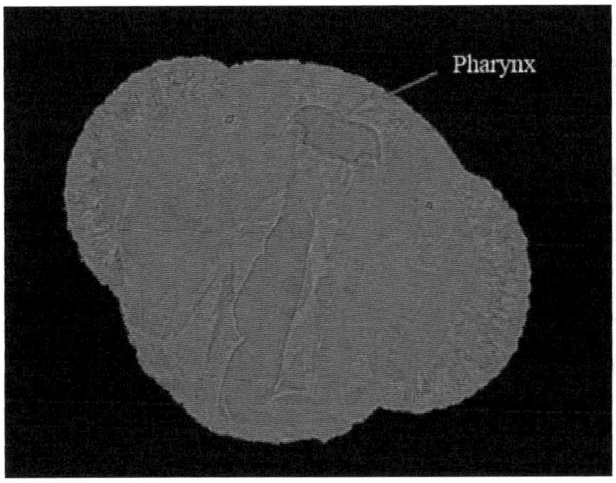

Figure 2. 3D rendering of the head of *R. prolixus* sacrificed 10 days after feeding.

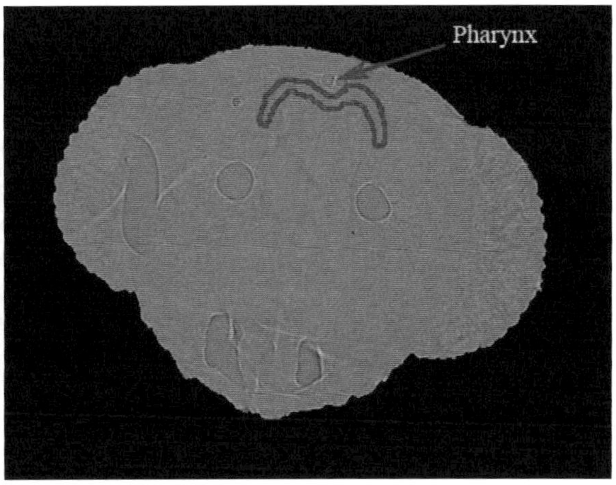

Figure 3. 3D rendering of the head of *R. prolixus* sacrificed 4 days after feeding.

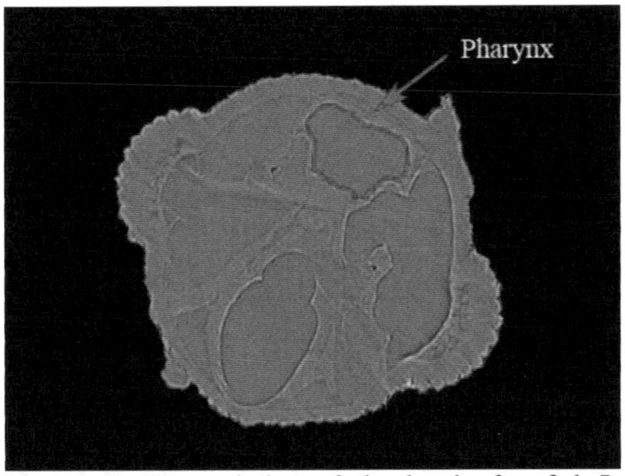

Figure 4. 3D rendering of the head of unfed *R. prolixus*.

Comparing this images one can realize that the muscles of unfed *R. prolixus* (figure 5) are thinner than the muscles in figures 3 and 4. Pharynx in unfed insects seems more dilated than in insects that were sacrificed 10 days after feeding.

In figures 5 and 6 it can be seen a virtual thin section of the head of *Rhodnius prolixus* obtained previously with the conventional setup at SYRMEP beamline {ARTIGOS ALMEIDA]. Comparing with figures 3 and 4 it is easy to realize that figures 6 and 7 do not show details seen in figures 2 and 3. The virtual thin sections are approximately in the same region.

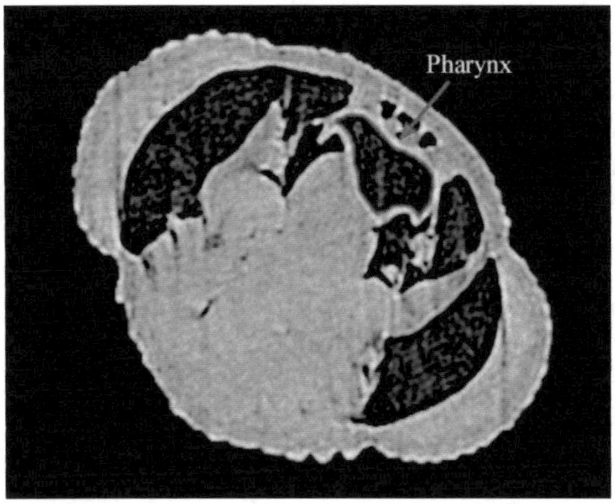

Figure 5. 3D rendering of the head of *R. prolixus* sacrificed 10 days after feeding.

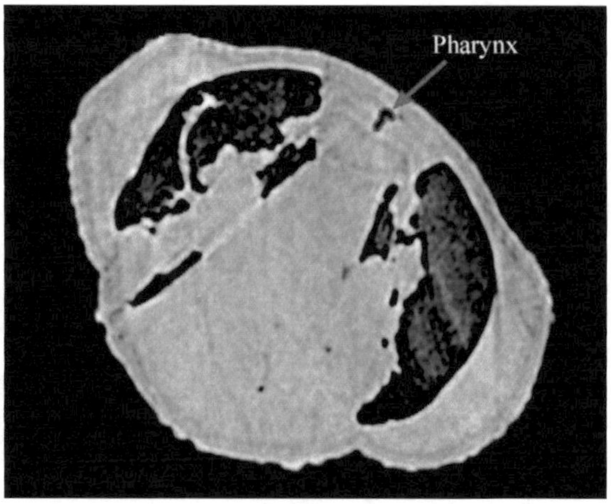

Figure 6. 3D rendering of the head of *R. prolixus* sacrificed 4 days after feeding.

External morphological features such as individual facets of the composed eyes and the detailed rugose cuticular features of the head capsule could be clearly visualized in figure 7.

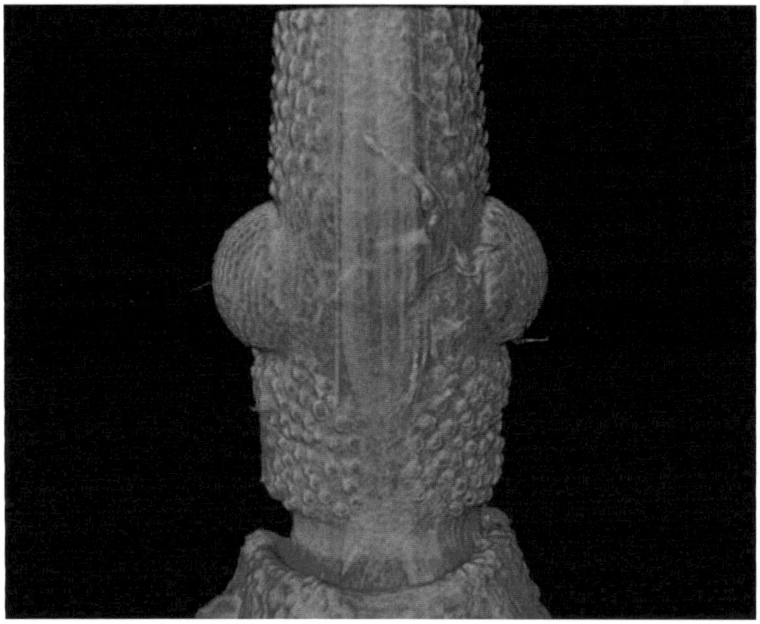

Figure 7. Volume-rendered SR-PhC-μCT image of *R. Prolixus*.

4. Conclusions

Classical anatomical techniques are costly in terms of time and effort and usually involve the destruction of the specimen under study. The SR-PhC-μCT technique enables an incomparable visualization of internal structures and is a non-destructive technique.

There is not any available vaccine to prevent from Chagas disease, so depending on the region, key tools are vector control, house improvement such as plastered walls, cement floors, corrugated-iron roofs and personal preventive measures. Anyway, controlling the transmission of *Trypanosoma cruzi* by means of its vector insect is much more efficient. In previous studies it has been shown that the control of *Trypanosoma cruzi* relies on the study of the head of *R. prolixus* and its structures. So, high resolution images from internal structures of the head of *R. prolixus* could be of great importance for biology and medicine to understand the main structures where *Trypanosoma cruzi* passes by.

Acknowledgments
We warmly thank Nicola Sodini from the SYRMEP beamline for their invaluable help during the experiment. These images were obtained during a visit (R.C.B.) within the Associate Scheme of the Abdus Salam International Center for Theoretical Physics (ICTP).

References
[1] Mizutania R, Suzuki Y 2012 Micron 43 104-115
[2] Hornschemeyer T, Goebbels J, Weidemann G 2006 J. Entomol 103 409-423
[3] Beutel R G, Ge S, Ho T, Hornschemeyeret T 2008 Cladistics 24 270-298
[4] Eberhard M J B, Lang D, Metscheret B 2010 Arthropod Struct Dev 39 230-241
[5] Zhang K, Li D, Zhu P 2010 Bioanal Chem 397 2143-2148
[6] Jakubeka J, Granja C, Damme J 2007 Nucl. Instrum. Meth. A 571 69–72
[7] Westneat M W, Socha J J, Lee W K 2008 Annu. Rev. Physiol. 70 119–142
[8] Heethoff M, Norton R A 2009 Invertebr. Biol. 128 332–339
[9] Betz O, Wegst U, Weide D 2007 J. Microsc. 227 51-71
[10] WHO-World Health Organization 2002 Control of Chagas Disease
[11] Almeida A P, Braz D, Nogueira L P, Colaço M V,SoaresJ, Cardoso S C,Garcia E S,Azambuja P, Gonzalez M S, Mohammadi S, Tromba G and Barroso R C 2012 JINST8C07004
[12] Almeida A P, Braz D, Nogueira L P, Colaço M V,SoaresJ, Cardoso S C,Garcia E S,Azambuja P, Gonzalez M S, Mohammadi S, Tromba G and Barroso R C 2013 Radiation Physics and Chemistry IN PRESS
[13] Abrami A, Arfelli F, Barroso R C 2005 Nucl. Instrum. Meth. A 548 221-227
[14] Garcia E S, Azambuja P, Forster H, Rembold H 1984 Nathurforsch 29 1155-1158
[15] Rembold H, Garcia E S 1989 Naturwissenschaften 76 77-78
[16] Garcia E S, Azambuja P, Forster H, Rembold H 1984 Nathurforsch 29 1155-1158
[17] Gonzalez M S, Nogueira N F S, Mello C B, Souza W, Schaub G A, Azambuja P, Garcia E S 1999 Experimental Parasitology 92 100-108

Journal of Physics: Conference Series **499** (2014) 012019 doi:10.1088/1742-6596/499/1/012019

Synchrotron radiation X-ray microtomography and histomorphometry for evaluation of chemotherapy effects in trabecular bone structure

R Alessio[1], L P Nogueira[2], A P Almeida[1], M V Colaço[2], D Braz[1], C B V Andrade[3], C Salata[3], S C Ferreira-Machado[4], C E de Almeida[3], G Tromba[5] and R C Barroso[2]

[1]COPPE - Universidade Federal do Rio de Janeiro, P.O. Box 68509, 21945-970, Rio de Janeiro, Brasil.
[2]Laboratório de Física Médica – Universidade do Estado do Rio de Janeiro, Rua São Francisco Xavier, 524 PJLF, sl 3007F, Rio de Janeiro, Brasil.
[3]Laboratório de Ciências Radiológicas – Universidade do Estado do Rio de Janeiro, Rua São Francisco Xavier, 524 PHLC, sl 136, Rio de Janeiro, Brasil.
[4]Departamento de Biologia Geral, Universidade Federal Fluminense, Outeiro de São João Batista, s/n, Niterói, Brasil
[5]Sincrotrone Trieste ScpA, Strada Statale S.S. 14 km 163.5, 34012 Basovizza, Trieste, Italy

E-mail: liebertrj@gmail.com

Abstract. Three-dimensional microtomography has the potential to examine complete bones of small laboratory animals with very high resolution in a non-invasive way. One of the side effects caused by some chemotherapy drugs is the induction of amenorrhea, temporary or not, in premenopausal women, with a consequent decrease in estrogen production, which can lead to bone changes. In the present work, the femur heads of rats treated with chemotherapy drugs were evaluated by 3D histomorphometry using synchrotron radiation microcomputed tomography. Control animals were also evaluated for comparison. The 3D tomographic images were obtained at the SYRMEP (SYnchrotron Radiation for MEdical Physics) beamline at the Elettra Synchrotron Laboratory in Trieste, Italy. Results showed significant differences in morphometric parameters measured from the 3D images of femur heads of rats in both analyzed groups.

1. Introduction

Chemotherapy often causes significant bone loss, marrow adiposity and haematopoietic defects. One of the side effects caused by some chemotherapy drugs is the induction of amenorrhea, temporary or not, in premenopausal women, with a consequent decrease in estrogen production. It leads to bone changes, similar to those presented in osteoporosis [1].

The osteoporosis is defined as a systemic skeletal disease characterized by low bone mass, micro-architectural deterioration and changes of bone tissue, resulting in increased bone fragility and susceptibility to fracture risk. It is generally accepted that trabecular bone strength depends not only on

Journal of Physics: Conference Series **499** (2014) 012019 doi:10.1088/1742-6596/499/1/012019

bone volume but also on its structures, which consist of connected bony plates. The investigation of a rat model for osteoporosis from X-ray microtomography has already been described previously [2, 3, 4].

Imaging techniques based on microtomography (μCT) enable three-dimensional (3D) non-destructive analysis of bone microarchitecture. μCT technique can be used to image and to quantify trabecular bone and this quantification has the capability to address the role of trabecular architecture on the mechanical properties of bone. The use of rat requires even higher spatial resolution, just because it has thinner bone structures. Trabecular diameter in rats is less than 100 μm while in humans, greater than 120 μm. To this goal, the combination of synchrotron radiation (SR) with μCT technique yields notable advantages, including high spatial resolution and coherence what can be achieved due to the high intensity of synchrotron facility and the natural collimation of the beam, and the wide spectral bandwidth allows continuous tuning of energy from a few keV to several tens of keV [5, 6].

For quantification, a volume of interest (VOI) has to be chosen in the way that it is comprised only of trabecular bones. For this VOI, the following parameters can be computed: bone volume-total tissue volume ratio (BV/TV), trabecular thickness (Tb.Th (mm)), trabecular separation (Tb.Sp (mm)), trabecular number (Tb.N (mm^{-1})) and structure model index (SMI). Different parameterization methods may be applied to extract quantitative architecture parameters from the sample such as in conventional histology. But this technique is somehow tedious and time-consuming, as it includes sample sectioning (the sample is destroyed) and the parameters are visually assessed in two dimensions, so the third dimension has to be added using the basis of stereology. In 3D computed microtomography, the histomorphometric parameters are assessed from 3D images of the bone structures, non-destructively, fastly, and precisely [7, 8, 9]. This technique was pioneered by Feldkamp and co-workers [10].

This work aims to evaluate changes in trabecular bone architecture of rats treated with a combination of chemotherapy drugs, using 3D computed microtomography and histomorphometry.

2. Materials and methods

2.1. *Specimens*

In this study, the experimental animals were ten adult female Wistar rats at 90 days of age, divided randomly into two groups, with n = 5 each. The treated group received doses of docetaxel and cyclophosphamide drugs (G1) while control group (G0) was sham treated. At the beginning of the study, the animals weighed an average of 200 g. The rats were acclimatized with standard conditions of temperature (25°**C**) and light-controlled environment (12 h light-dark cycles). Food and water were provided *ad libitum*. The treatment lasted for 1 month in cycles of 4 and 7 day of intervals between them. At the same time, untreated animals of the same age were sham treated as control. The rats were sacrificed by direct heart KCl injection at 150 days post-treatment, at 240 days of age, and femurs were excised, cleaned and let to air dry for at least 72 hours. Ethics permission to utilize the animals for the research described in this paper was obtained from Ethics Committee on Animal Research of the State University of Rio de Janeiro (Process CEA/010/2012). Figure 1 illustrates the location from where images were acquired.

Journal of Physics: Conference Series **499** (2014) 012019 doi:10.1088/1742-6596/499/1/012019

2.2. Synchrotron radiation computed microtomography (SR-μCT)

All the specimens were imaged using the new high resolution microCT setup which has been recently available at the Synchrotron Radiation for Medical Physics (SYRMEP) beamline of the Elettra Synchrotron Light Laboratory (Trieste, Italy). The SYRMEP light source is one of the bending magnets of ELETTRA. The horizontal acceptance covered by the front-end light-port is 7 mrad. The useful energy range is 8.5-35 keV. Typical flux measured at the sample position at 17 keV is about $1.6x10^8$ ph/mm^{2}s, with a stored electron beam of 300 mA when ELETTRA is operated at 2 GeV, while it is $5.9x10^8$ ph/mm^{2}s, with 140 mA at 2.4 GeV. A custom-built ionization chamber, placed upstream to the sample, is used to determine the exposure on the sample, and hence to calculate the delivered dose [11]. In order to optimize the performances of the μCT setup for high resolution experiments in the SYRMEP beamline, 25-μm thick single crystal of cerium-doped lutetium–aluminum garnet (Lu3Al5O12) scintillator screen (Crytur, Czech Republic) was coupled to an air-cooled 16 bit CCD camera (Photonic Science, KAI 4022M CCD, 2048 x 2048 full frame) via a visible light microscope optics (LEICA), effective pixel size of 1.03 μm. This system designed to achieve up to 2 μm spatial resolution was used in white/pink X-ray beam mode that provides a nearly parallel, laminar-section X-ray beam with a maximum area of 100 mm (horizontal) x 6 mm (vertical) at a distance of about 15 m from the source.

For each bone sample, 1800 radiographic images were acquired over an angular range of 360° with angular step of 0.2°. Sample-detector distance was set to 9 cm and scanning time was approximately 1h for each sample. The sample was positioned so that the region of femur head lied on the field of view of the detector, not been necessary to cut the sample. Palladium (0.047 mm) and Silicon (1.5 mm) filters were used to cut the low energy X-ray components therefore, the average energy was around 24 keV. The 2D radiographies are normalized by using flat (images without the samples) and dark (background) images. This procedure allows one to take into account incident beam non-uniformities and to correct fixed noise due to the efficiency of the detector elements. After radiographic data acquisition, the SYRMEP Tomo Project software was used to reconstruct the slices [12]. The reconstruction was performed using filtered back projection with Shepp Logan filter. The quantitative analysis was performed in a VOI with 200 x 158 x 500 pixels, starting 0.5 mm below the top of femur head. The ImageJ® software [13] was used to render the 3D volumes and BoneJ software (ImageJ plugin) [14] was employed for quantifying the samples.

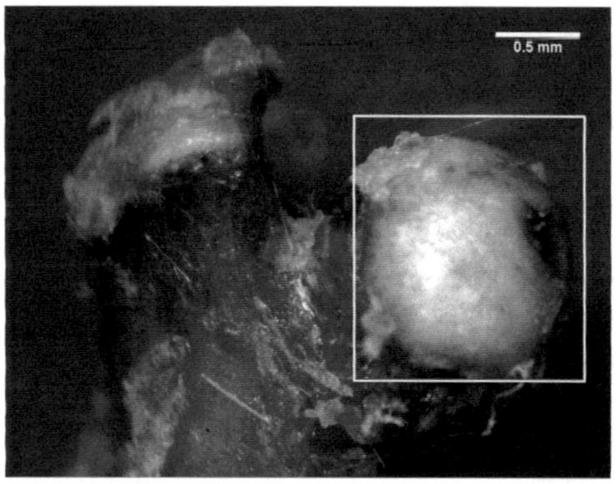

Figure 1. Femur head - location from where the microtomographic images were obtained.

2.3. Morphologic parameters

Prior to quantification of morphologic parameters, the images had to be binarized so that voxels corresponding to bone could be distinguished from those of background. To distinguish bone tissue from marrow and background, an optimal threshold for each specimen was determined. The method used is implemented in ImageJ, called IsoData, also known as Iterative Intermeans [15].

The morphologic parameter BV/TV was calculated by the number of foreground (bone) voxels divided by the total number of voxels in the image. The 3D model-independent parameter Tb.Th was computed based on the calculation of the local thickness volume [16]. The local thickness $\tau(n)$ at any point $n \in \Omega \subset R^3$ is defined as the diameter of the largest sphere containing the point n and which is completely inside the structure Ω, ie:

$$\tau(n) = 2\max\{r / n \in sph(x,r) \subseteq \Omega, x \in \Omega\} \tag{1}$$

This definition was given in the context of the continuous space utilizing the Euclidean distance. The trabecular separation Tb.Sp was computed using the same process but applied to the complement of the binary trabecular bone structure. The number of trabeculae, Tb.N, was determined using:

$$Tb.N = \frac{BS}{2TV} \tag{2}$$

BS, the surface of trabecular bone, was calculated using a triangular surface mesh by marching cubes and bone surface area can be assumed to be the sum of the areas of the triangles making up the mesh [17].

An estimation of the plate-rod characteristic of the structure is achieved using the Structure Model Index, SMI [18]. This parameter is calculated by a differential analysis of a triangulated surface of a structure and is defined as

$$SMI = 6\,.\frac{BV\dfrac{dBS}{dr}}{BS^2} \tag{3}$$

where dBS/dr is the surface area derivative with respect to a linear measure r, relative to the half-thickness. For an ideal plate and rod structure the SMI value is 0 and 3, respectively. For structure with both plates and rods of equal thickness the value is between 0 and 3, depending on the volume ratio between rods and plates.

3. Results

The femoral head samples were scanned to reveal the two-dimensional (2D) and three-dimensional (3D) trabecular microstructure. Figure 2 presents the 2D and 3D microstructure of a typical specimen from G0 and G1 groups. The 3D rendered images, together with 2D sections of the G1 group (Figure 2) reveal the possible changes in the microarchitecture of cancellous femoral head. It can be seen that the microstructure of femoral head in the chemotherapy group showed a greater spacing between the trabeculae and loss of interconnections, compared with control group.

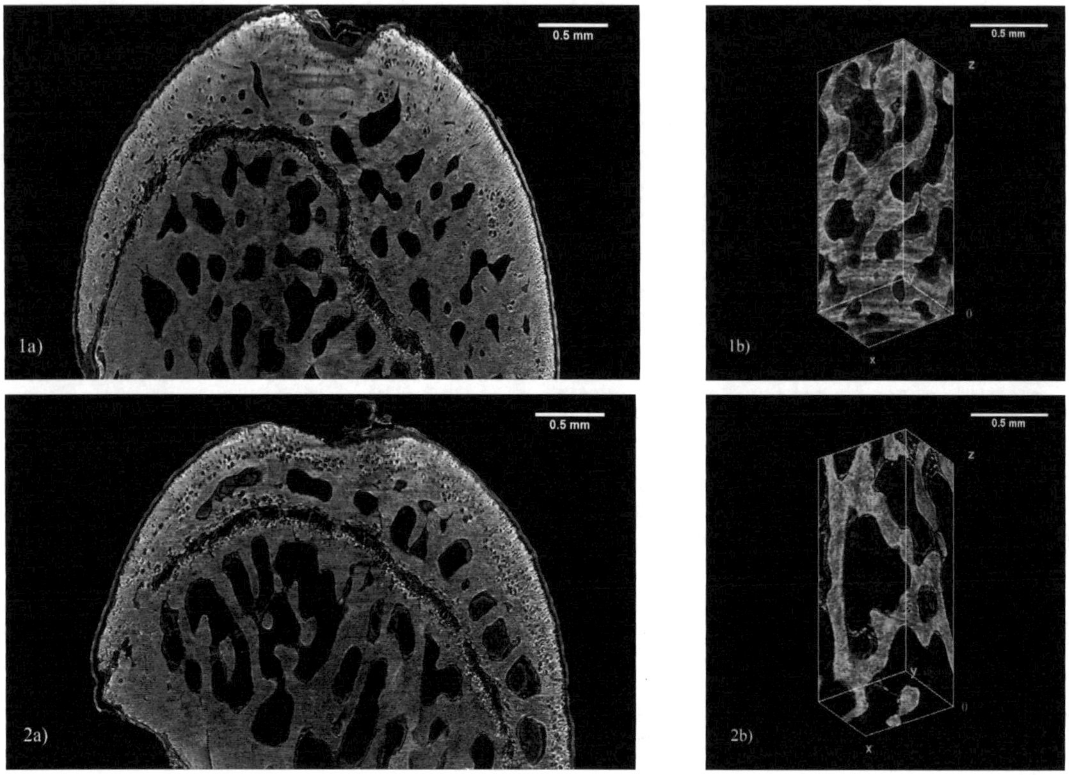

Figure 2. Two-dimensional trabecular microstructure from SR-μCT images of rat femoral head (a) and 3D ROI from where histomorphometry was computed: (1) Control group and (2) Chemotherapy group.

On the original 3D image, morphometric indices were determined directly from the binarized volume of interest (VOI). Statistical differences among groups were evaluated with Student's t-test. P-values less than 0.05 were considered to be significant. Table 1 indicates the mean values and respective standard deviations of the different trabecular parameters for both groups. The statistical significance between the two groups is reported in the last line through the p-value.

The results reveal a significant decrease in BV/TV as well as an increase in spacing (Tb.Sp) but not significant for rats treated with chemotherapic drugs. This fact can be clearly seen in microtomographic images of the samples of this group (Figure 2). The trabecular thickness (Tb.Th) increased 7% in the G1 group, compared to the control one, but not significantly. The number of trabeculae per millimeter (Tb.N) decreased of 13% in G1 group compared with control one (P = 0.095). Bone volume to total volume ratio (BV/TV) decreased 31% (P = 0.006) with a consequent increase of 36% (P = 0.063) in trabecular spacing (Tb.Sp). The Structure Model Index (SMI) increased significantly 154% (P = 0.02) in G1 group comparing to G0, showing values varying between 0.244 and 2.240, indicating that trabeculae appear to have mixed plate-rod structures. The lowest value obtained (0.244) is closer to the plate-like, while the highest value (2.240) resembles rod-like structures. The deterioration of cancellous bone structure due to aging and diseases such as osteoporosis is characterized by a conversion from plate elements to rod elements. Consequently the terms "rod-like" and "plate-like" are frequently used for a subjective classification of cancellous bone [19]. Group treated with chemotherapy presented SMI of 1.840, indicating that structures are closer to rod-like, while control group presented SMI of 0.724, meaning more plate-like structures which reflects higher mechanical strength.

Table 1. Histomorphometric indices for femur head.

Groups	BV/TV	Tb.Th (mm)	Tb.Sp (mm)	Tb.N (mm^{-1})	SMI
G0 (n = 5)	0.494 ± 0.041	0.028 ± 0.002	0.076 ± 0.019	8.454 ± 0.880	0.724 ± 0.480
G1 (n = 5)	0.339 ± 0.040	0.030 ± 0.005	0.103 ± 0.021	7.379 ± 0.921	1.840 ± 0.598
P value	P = 0.006[a]	P = 0.460	P = 0.063	P = 0.095	P= 0.02[a]

[a] Significantly different ($P < 0.05$)

4. Conclusions

The use of synchrotron sources with considerably higher flux opens up new possibilities in the analysis of 3D images from bone samples, in particular to access quantitative analysis of trabecular microarchitecture. Three-dimensional SR-μCT was used to investigate bone architecture in femur head site. After image acquisition, 3D image processing technique was used to fully exploit 3D data. The advance in three-dimensional imaging techniques allows studying the organization of trabeculae in great detail. Concerning the obtained results for control and treated bones from same skeletal site, a certain declining bone volume fraction was achieved. The relative bone loss (decreased BV/TV) between chemotherapy group and control group in the femur head accompanied by lower Tb.N and higher Tb.Sp seems to indicate that chemotherapy induced bone loss and it is manifested through decreased connectivity and loss of thin trabecular elements. SMI seems to indicate that structures are tending to turn from plate to rod-like as chemotherapy is applied, suggesting a lower mechanical strength in the bones of chemotherapy treated animals.

In conclusion, the results obtained could be used in forming the basis for comparison of the bone microarchitecture and the resulting 3D high resolution images along with histomorphometric quantification can be a valuable tool for predicting bone fragility.

Acknowledgments

The authors thank the Brazilian agencies CNPq (Conselho Nacional de Desenvolvimento Científico e Tecnológico), FAPERJ (Fundação de Amparo à Pesquisa do Estado do Rio de Janeiro) and CAPES (Coordenação de Aperfeiçoamento de Pessoal de Nível Superior) for financial support. This work was supported by ICTP-ELETTRA Users Programme (Projects SYRMEP 2007834 and 20090192). These images were obtained during a visit (R.C.B.) within the Associate Scheme of the Abdus Salam International Center for Theoretical Physics (ICTP). We acknowledge the SYRMEP group for help during image acquisition and reconstruction.

References

[1] Marcus R, Feldman D, Dorothy N and Clifford J 2008 *Osteoporosis* **2** 1337
[2] Barbier A and Martel C 1999 *J. Bone Miner Metab* **17** 37
[3] Laib A, Barou O, Vico L, Lafage-Proust M, Alexandre C and Ruegsegger P 2000 *Med Biol Engng Comput* **38** 326
[4] Martìn-Badosa E, Elmoutaquakkil A, Nuzzo S, Amblard D, Vico L, Peyrin F 2003 *Medical Imaging and Graphics* **27** 447
[5] Muller R 2009 *Nat. Rev. Rheumatol.* **5** 373

[6] Margaritondo G and Meuli R 2003 *Eur. Radiol.* **13** 2633
[7] Nogueira L, Almeida A P, Braz D, Andrade C B, Salata C, Tromba G, Almeida C E and Barrroso R C 2012 *Appl. Radiat. Isot.* **70** 1296
[8] Nogueira L, Almeida A P, Braz D, Andrade C B, Tromba G, Almeida C E and Barrroso R C 2011 *AIP Conf. Proc.* **39** 1437
[9] Nogueira L, Braz D, Oliveira L F, Pinheiro C J G, Dreossi D, Tromba G and Barroso R C 2010 *Micron* **41** 990
[10] Feldkamp L A, Goldstein S A, Parfitt M A, Jesion G and Kleerekoper M 1989 *J. Bone Miner. Res.* **4** 3
[11] Abrami A, Arfelli F, Barroso R C, Bergamaschi A, Billè F, Bregant P, Brizzi F, Casarin K, Castelli E, Chenda V, Dalla Palma L, Dreossi D, Fava C, Longo R, Mancini L, Menk R H, Montanari F, Olivo A, Pani S, Pillon A, Quai E, Ren Kaiser S, Rigon L, Rokvic T, Zanetti A and Zanini F 2005 *Nuclear Instruments & Methods in Physics Research A* **548** 221.
[12] Montanari F 2003 http://www.ts.infn.it/physics/experiments/syrma/SYRMEP/Tutorial tomo program.ppt
[13] Schneider A, Rasband W S and Eliceiri K W 2012 *Nat. Meth.* **9** 671
[14] Doube M, Kłosowski M M, Arganda-Carreras I, Cordelieres F, Dougherty R P and Jackson. J 2010 *Bone.* **47** 1076
[15] Riddler W and Calvard S 1978 *IEEE Tran. Syst. Man. Cybern* **8** 630
[16] Hildebrand T and Ruegsegger P 1997 *J Microsc.* **185** 67
[17] Lorensen W E and Cline H E *Comput Graph (SIGGRAPH 87 Proceedings)* 1987 **21** 163
[18] Hildebrand T and Ruegsegger P 1997 *Comput Meth Biomed Engng* **1** 15
[19] Ding M and Hvid I 2000 *Bone* **26** 291

Journal of Physics: Conference Series **499** (2014) 012020 doi:10.1088/1742-6596/499/1/012020

Ptychographical imaging of the phase vortices in the x-ray beam formed by nanofocusing lenses

D Dzhigaev[1,2], U Lorenz[1,*], R P Kurta[1], F Seiboth[3], T Stankevic[4], S Mickevicius[4], A Singer[1,†], A Shabalin[1], O M Yefanov[5], M N Strikhanov[2], G Falkenberg[1], C G Schroer[3], R Feidenhans'l[4] and I A Vartanyants[1,2]

[1] Deutsches Elektronen-Synchrotron DESY, Notkestraße 85, D-22607 Hamburg, Germany
[2] National Research Nuclear University, "MEPhI", 115409 Moscow, Russia
[3] Institute for Structural Physics, Technische Universität Dresden, D-01062 Dresden, Germany
[4] Niels Bohr Institute, University of Copenhagen, DK-2100 Copenhagen, Denmark
[5] Center for Free-Electron Laser Science CFEL, Notkestraße 85, D-22607 Hamburg, Germany

E-mail: `dmitry.dzhigaev@desy.de`

Abstract. We present the ptychographical reconstruction of the x-ray beam formed by nanofocusing lenses (NFLs) containing a number of phase singularities (vortices) in the vicinity of the focal plane. As a test object Siemens star pattern was used with the finest features of 50 nm for ptychographical measurements. The extended ptychographical iterative engine (ePIE) algorithm was applied to retrieve both complex illumination and object functions from the set of diffraction patterns. The reconstruction revealed the focus size of 91.4±1.1 nm in horizontal and 70±0.3 nm in vertical direction at full width at half maximum (FWHM). The complex probe function was propagated along the optical axis of the beam revealing the evolution of the phase singularities.

1. Introduction

Phase singularities are a common feature of different forms of waves and represent a fundamental topology property of wave fields [1]. In 1804 Young has described effects of interference from different types of obstacles in the path of the light beam [2]. When three or more waves interfere, points of zero intensity could appear. At these positions the phase is undefined (singular), and, in general, all phase values in the interval $[0; 2\pi]$ occur around a vortex point, leading to a circulation of the optical energy. Phase singularities were discussed in the terms of dislocations in wave trains in 1974 [3] and were first observed in the optics of visible light [4]. Ten years ago phase vortices were observed in the x-ray regime using spiral zone plate at 9 keV photon energy [5]. Phase singularities could also appear after interaction with the crystal lattice dislocations [6, 7].

One generally describes light as a plane wave, that is an electromagnetic field with a constant phase $\psi(\mathbf{r})=\rho(\mathbf{r})\exp(i\varphi(\mathbf{r}))$, where $\rho(\mathbf{r})$ is the real amplitude and $\varphi(\mathbf{r})$ is the phase at the

* Present address: Institut für Chemie, Universität Potsdam, Karl-Liebknecht-Straße 24-25, D-14476 Potsdam, Germany
† Present address: The University of California, San Diego, La Jolla, CA 92093, USA

Journal of Physics: Conference Series **499** (2014) 012020 doi:10.1088/1742-6596/499/1/012020

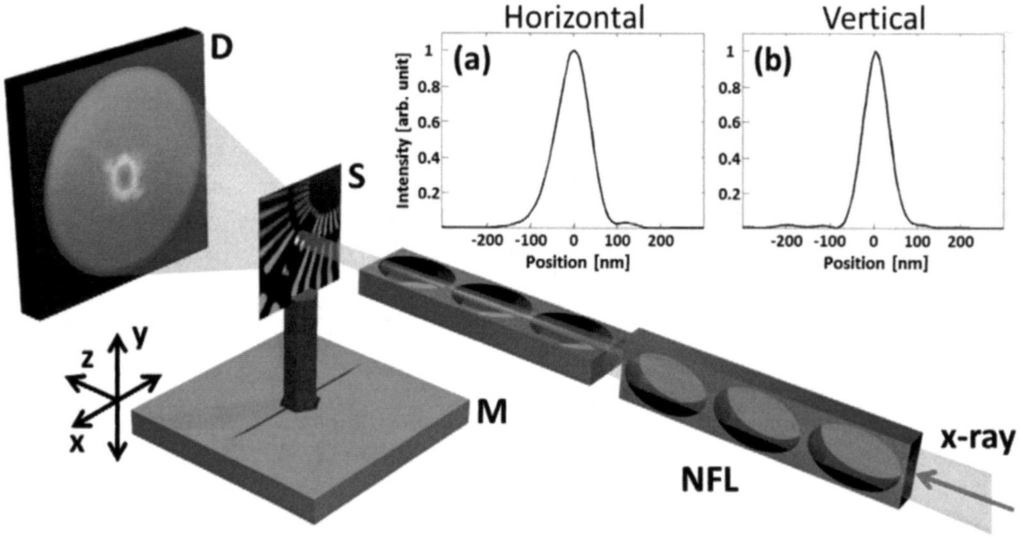

Figure 1. Experimental setup. The incoming x-ray beam (red arrow) goes along the Z axis and is focused by a pair of perpendicularly positioned NFLs. The test sample (S) in the form of the Siemens star is mounted on the movable stage (M) and is illuminated at the positions of a raster grid. Diffraction patterns are collected by the detector (D) 2.1 m downstream. Insets (a) and (b) show horizontal and vertical profiles of the normalized intensity of the reconstructed probe function across the central maximum.

position **r**. This field extends infinitely and normally to the Poynting vector. Vortices can be characterized by the integer number S (positive or negative) that is called a strength or topological charge of the singularity and is determined by $S=(1/2\pi)\oint_C d\varphi$, where C is any closed path around the vortex point. Dynamics of the vortices include processes of nucleation, annihilation, and propagation in three dimensions [8].

Ptychography, first proposed by Hoppe in the field of electron microscopy more than 40 years ago [9], became a well established x-ray microscopy technique during the last decade [10, 11]. Development of the phase retrieval algorithms [12] made this approach especially useful for the beam characterization with the use of test patterns [13]. It allows to reconstruct the complex illumination function (probe) and the complex object function simultaneously. In this work we apply ptychography technique to reconstruct the wave field generated by nanofocusing lenses (NFLs) in the hard x-ray regime.

2. Experiment

We performed an experiment at the nanoprobe end station of P06 beamline at PETRA III synchrotron source at DESY [14]. The geometry of the experiment is shown in Fig. 1. Two perpendicularly positioned NFLs based on parabolic compound refractive x-ray lenses (CRLs) were used to obtain a nano-sized focus of the incident x-ray beam with 15.25 keV energy. These lenses were fabricated by electron beam lithography and deep trench reactive etching. The lenses were produced from silicon, because they can be shaped accurately on the sub-micrometer scale [15]. The flux of the beam in the focus was 4×10^7 photons/sec. A Pilatus 300K hybrid-pixel detector (Dectris, Switzerland) with the pixel size of the $172 \times 172\ \mu m^2$ was used. The detector was positioned 2.1 m downstream from the sample. A tantalum test sample in the form of a Siemens star, fabricated by nanolithography was mounted on the movable sample stage and

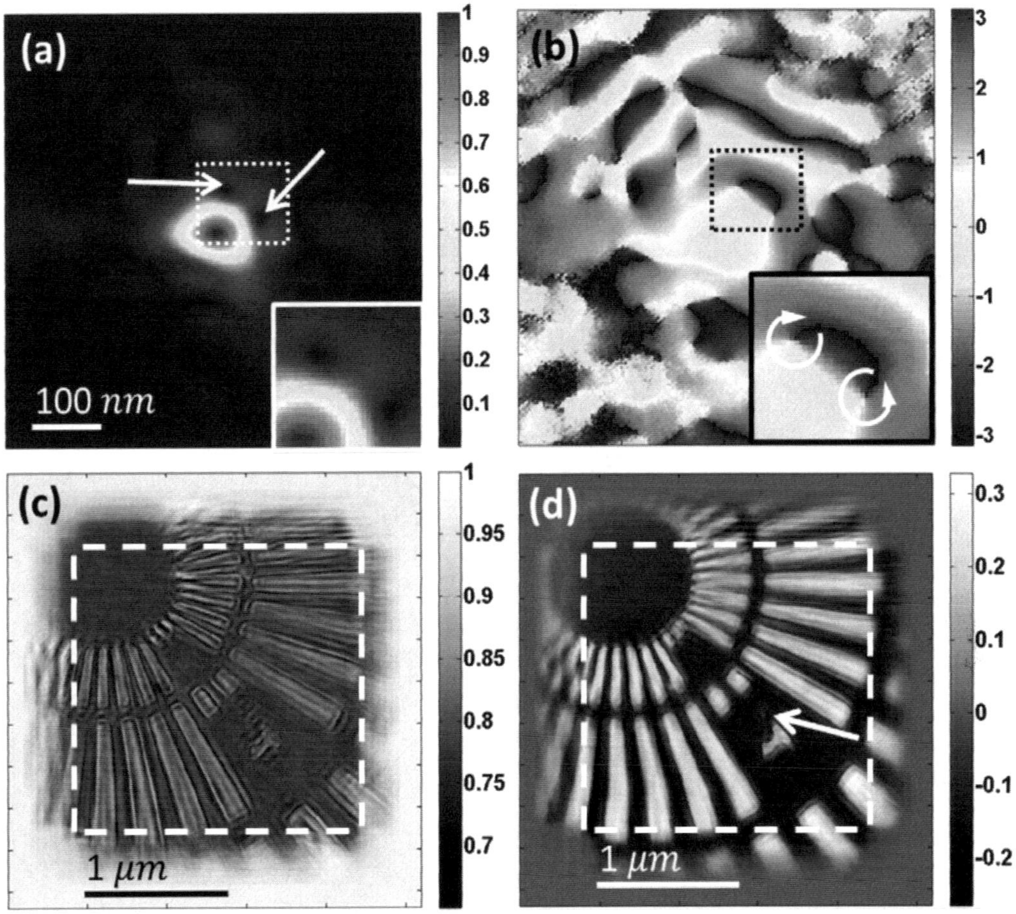

Figure 2. Results of ptychographical reconstruction at the sample plane. (a) Amplitude of the probe function. White arrows indicate two points of zero intensity corresponding to phase singularities (see inset for an enlarged view). (b) Phase of the probe function. In the inset an enlarged view of two singularities with opposite directions is shown. (c) Amplitude of the object function. (d) Phase of the object function. Smallest resolved feature is a dot that is 20 nm in diameter. It is shown by an arrow. The color bars in (a) and (c) show normalized values of the amplitude functions and in (b) and (d) values of the phase in radians. Region of the scan is outlined by white dashed lines in (c) and (d).

positioned in the focal plane. The ptychographical scan was performed on a Cartesian grid with 50 nm step size and 41×41 scan positions, in the horizontal and vertical directions, perpendicular to the optical axis of the beam. The step size corresponds to 58% of the probe overlap [16]. The acquisition time was 0.3 seconds per scan position.

3. Results and discussion

An extended ptychographical iterative engine (ePIE) algorithm [17] was applied to determine the complex probe (see Fig. 2 (a, b)) and complex object function (see Fig. 2 (c, d)). They were reconstructed from 1681 diffraction patterns. The field of view was about 2×2 μm^2 and is shown by white dashed line in Fig. 2 (c, d). Reconstruction procedure started from random initial guess of the probe function and a uniform object with a constant transmissivity without a phase

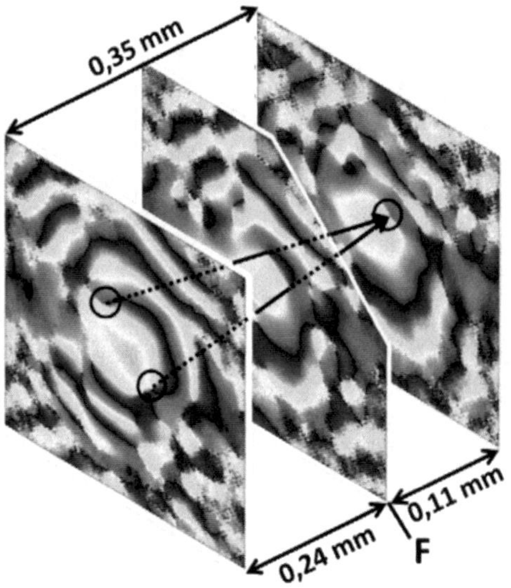

Figure 3. Propagation of the wave field. Three 2D cuts perpendicular to the beam propagation direction are shown: 0.24 mm in front of the focal plane (F), at the focal plane, and 0.11 mm behind the focal plane. At the first position the pair of vortices nucleate, at the last one they annihilate.

shift. The final result was obtained after 100 iterations. The pixel size in this reconstruction is 6 nm. The smallest detail in the object pattern that was resolved is 20 nm in size (see Fig. 2 (c, d)). Distortions in the reconstructed image are due to the drifts of the vertical motors during measurements.

The size of the illumination spot on the sample was obtained by fitting of the intensity distribution by a Gaussian function. It turned out to be 91.4±1.1 nm in horizontal and 70±0.3 nm in vertical direction at full width at half maximum (FWHM) (see insets in Fig. 1). In the reconstructed images of the probe function (see Fig. 2 (a, b)) one can clearly see a pair of zero amplitude regions and corresponding singularities in the phase. In the inset of Fig. 2 (b) the direction of the phase change is shown by the white arrows. The phase vortices determined by ptychographical imaging correspond to the S={1,-1} topological charges.

The reconstructed complex wave field profile was propagated from 1 mm in front of the focus to 1 mm behind it in the frame of paraxial approximation. The calculation was performed by using numerical implementation of the Fresnel near field propagation equation [18]. The region, where the pair of vortices nucleate and annihilate, is shown in Fig. 3. In three dimensions the wave function $\psi(\mathbf{r})$ vanishes at each point of the line, called nodal line. These lines are clearly seen in two dimensional cuts through the propagated wave field at different planes perpendicular to the beam propagation direction (see Fig. 3). The length of the vortex lines is 0.24 mm before and 0.11 mm after the focal plane, giving in total 0.35 mm.

The presence of vortices in the illuminating wave field may cause degradation of the quality of the phase retrieval procedure. In this case, a number of corrections should be taken into account during reconstruction as shown for high-resolution transmission electron microscopy (TEM) [19] and classical wave fields [20]. Nevertheless, the ptychography technique is very tolerant to imperfections of the probe and the features of the object much smaller than the beam size can be reconstructed as demonstrated in this work.

4. Conclusions

We have obtained the ptychographical reconstruction of the x-ray field focused by NFLs containing a number of phase singularities (vortices) in the vicinity of the focal plane. Siemens star pattern was used as a test object with the finest feature of 20 nm. After inversion procedure with the ePIE algorithm a complex wave field function was obtained. The reconstruction revealed the focus size (FWHM) of 91.4±1.1 nm in horizontal and 70±0.3 nm in vertical direction. The pair of vortices closest to the central maximum of the beam with the topological charges ±1 were propagated from nucleation to annihilation plane. The length of the nodal line for these singularities was 0.35 mm in total. Appearance of the vortices in the focal region of the nano-focused beam could possibly affect the quality of the phase retrieval procedure and should be taken into account in future work.

Acknowledgments

We acknowledge fruitful discussions and support of this project by E. Weckert, and careful reading of the manuscript by M. Sprung. This work was supported by the EU grant for the project 280773 “Nanowires for solid state lighting” , the Virtual Institute VH-VI-403 of the Helmholtz Association and by BMBF Proposal 05K10CHG “Coherent Diffraction Imaging and Scattering of Ultrashort Coherent Pulses with Matter” in the framework of the German-Russian collaboration “Development and Use of Accelerator-Based Photon Sources”.

References

[1] Berry M V, Upstill C and Wolf E 1980 Progress in Optics **18** 359
[2] Dennis M R, O'Holleran K and Padgett M J 2009 Progress in Optics **53** 293
[3] Nye J F and Berry M V 1974 Proc. R. Soc. Lond. A. **336** 165
[4] Tamm Chr and Weiss C O 1990 JOSA B **7** 1034
[5] Peele A G, Philip J M, Paterson D, Chanh Q T, Mancuso A P, Nugent K A, Hayes J P, Harvey E, Lai B and Ian M 2002 Opt. Lett. **27** 1752
[6] Jacques V L R, Ravy S, Le Bolloc'h D, Pinsolle E, Sauvage-Simkin M and Livet F 2011 Phys. Rev. Lett. **106** 065502
[7] Takahashi Y, Suzuki A, Furutaku S, Yamauchi K, Kohmura Y and Ishikawa T 2013 Phys. Rev. B **87** 121201(R)
[8] Chen M and Roux F S 2008 J. Opt. Soc. Am. A **25** 1279
[9] Hoppe W 1969 Acta Cryst. A **25** 495
[10] Rodenburg J M, Hurst A C, Cullis A G, Dobson B R, Pfeiffer F, Bunk O, David C, Jefimovs K and Johnson I 2007 Phys. Rev. Lett. **98** 034801
[11] Thibault P, Dierolf M, Menzel A, Bunk O, David C and Pfeiffer F 2008 Science **321** 379
[12] Thibault P, Dierolf M, Bunk O, Menzel A and Pfeiffer F 2009 Ultramicroscopy **109** 338
[13] Schropp A, Boye P, Feldkamp J M, Hoppe R, Patommel J, Samberg D, Stephan S, Giewekemeyer K, Wilke R N, Salditt T, Gulden J, Mancuso A P, Vartanyants I A, Weckert E, Schöder S, Burghammer M and Schroer C G 2010 Appl. Phys. Lett. **96** 091102
[14] Schroer C G, Boye P, Feldkamp J M, Patommel J, Samberg D, Schropp A, Schwab A, Stephan S, Falkenberg G, Wellenreuther G and Reimers N 2010 Nucl. Instrum. Meth. A **616** 93
[15] Boye P 2009 TU Dresden Thesis
[16] Bunk O, Dierolf M, Kynde S, Johnson I, Martib O and Pfeiffer F 2008 Ultramicroscopy **108** 481
[17] Maiden A M and Rodenburg J M 2009 Ultramicroscopy **109** 1256
[18] Paganin D M Coherent X-Ray Optics 2006 Oxford Univ. Press Oxford (UK)
[19] Allen L J, Faulkner H M L, Oxley M P and Paganin D 2001 Ultramicroscopy **88** 85
[20] Allen L J, Faulkner H M L, Nugent K A, Oxley M P and Paganin D 2001 Phys. Rev. E **63** 037602

Journal of Physics: Conference Series **499** (2014) 012021 doi:10.1088/1742-6596/499/1/012021

Structural properties of π-π conjugated network in polymer thin films studied by x-ray cross-correlation analysis

R P Kurta[1], L Grodd[2], E Mikayelyan[2], O Y Gorobtsov[1,3], I Fratoddi[4], I Venditti[5], M Sprung[1], S Grigorian[2] and I A Vartanyants[1,6]

[1] Deutsches Elektronen-Synchrotron DESY, Notkestraße 85, D-22607 Hamburg, Germany
[2] Department of Physics, University of Siegen, Walter-Flex-Strasse 3, D-57072 Siegen, Germany
[3] National Research Center 'Kurchatov Institute', Kurchatov Square 1, 123182 Moscow, Russia
[4] Department of Chemistry and Center for Nanotechnology for Engineering (CNIS), University of Rome Sapienza, P.le A. Moro 5, I-00185 Rome, Italy
[5] Department of Chemistry, University of Rome Sapienza, P.le A. Moro 5, I-00185 Rome, Italy
[6] National Research Nuclear University, "MEPhI", 115409 Moscow, Russia

E-mail: ruslan.kurta@desy.de, grigorian@physik.uni-siegen.de

Abstract. We present results of an x-ray study of structural properties of π-π conjugated networks in semicrystalline polymer thin films using nanofocused x-ray beam. We applied the x-ray cross-correlation analysis to x-ray scattering data from blends of poly(3-hexylthiophene) (P3HT) with gold nanoparticles. The Fourier spectra of the intensity cross-correlation functions for different films contain non-zero components of orders $n = 2, 4$ and 6 revealing structural order in the system. The results suggest inhomogeneous structure of both samples, with preferential edge-on orientation of domains in one sample and mixed orientation in another sample.

1. Introduction

Charge carrier mobility is a decisive parameter of organic field-effect transistors (OFETs). It determines the overall device performance. Semicrystalline conjugated polymers are promising candidates for OFETs. Typical features of these polymers are mixtures of poor and well organized domains. Usually the well organized domains are addressed to be crystalline and assumed to strongly improve device performance. Among different conjugated polymers polythiophenes have received an increasing attention in recent years due to their attractiveness for OFETs and solar cell applications [1]. One of the most often used polythiophene is poly(3-hexylthiophene) (P3HT) which crystalline domain sizes can vary from tens to few hundreds of nanometers, depending on preparation techniques [2, 3].

A conventional technique for structural analysis of organic thin films is grazing incidence x-ray diffraction (GIXD) which, in many cases of pristine P3HT and its blends, reveals a preferential edge-on orientation [4]. Generally it is assumed that such orientation with π-π stacking being parallel to the sample surface enhances OFET performance [5]. Typically, the scattering signal from spin cast polythiophene films probed by GIXD is rather weak. This is especially true for the in-plane peaks associated with the π-π conjugated network, even measured at highly intense synchrotron sources. Moreover, structural information on the scale of individual polythiophene stacks can only be achieved by combination of GIXD analysis with complementary high resolution transmission electron microscopy (TEM) technique [6].

The novel x-ray cross-correlation analysis (XCCA) technique [7–12] is an attractive way of structural characterization of partially disordered samples that can provide complementary information. It is a powerful tool to determine order-disorder variations across the sample from coherent diffraction patterns. For example, nanobeam studies in transmission geometry on pristine P3HT films were successfully realized using XCCA, unraveling the local structural information of P3HT nanodomains hidden in the powder average [13]. Recently it was demonstrated a strong enhancement of the local order of conjugated network induced by additives [14]. In the current study we imbedded gold nanoparticles in a thin P3HT film with the purpose to enhance both electrical and structural properties of the host polymeric matrix.

The aim of current study is to analyze structural variations of P3HT host matrix upon small addition of gold nanoparticles stabilized with fluorene derivatives (AuNPs-SFL) from the coherent diffraction pattern by XCCA technique. A special goal of our work is to understand the structural properties of the π-π conjugated P3HT network on the nanoscale defined by a nanoprobe beam. This is likely to have a strong effect on the intermolecular charge transport and consequently on the electronic properties. Therefore x-ray nanobeam is an irreplaceable tool to observe local order and orientation of conjugated network induced by Au nanoparticles.

2. Theory

We consider coherent x-ray scattering from a partially ordered polymer/AuNPs blend. Schematic setup of x-ray transmission geometry is shown in Fig. 1(a) where the strongest intensity ring is associated with π-π conjugated network. One can expand the scattered intensity $I(\mathbf{q})$ at each momentum transfer vector $\mathbf{q} = (q, \varphi)$ into angular Fourier series as

$$I(q, \varphi) = I_0(q) + 2\sum_{n=1}^{\infty} I_n(q)\cos(n\varphi), \tag{1}$$

where q and φ are the radial and azimuthal coordinates of the vector $\mathbf{q}$ in polar coordinate system, and $I_n(q)$ are the Fourier components of intensity $I(q, \varphi)$. Depending on the system under investigation and experimental conditions, these complex Fourier components may contain rich information about the structure of a single particle in a disordered system [11], degree of bond-orientational order in liquid crystals [15] or local structure of amorphous system [7].

The angular XCCA enables direct determination of the Fourier components $I_n(q)$ from the ensemble of diffraction patterns measured at different positions on the sample [8–10]. For the i-th diffraction pattern a two-point cross-correlation function (CCF) can be defined as

$$C^{i,i}(q, \Delta) = \left\langle I^i(q, \varphi) I^i(q, \varphi + \Delta)\right\rangle_{\varphi}, \tag{2}$$

where Δ is the angular coordinate, $\langle \ldots \rangle_{\varphi}$ denotes the angular average of the scattered intensity around the ring of a radius q, and $I^i(q, \varphi)$ indicate the intensities measured on the same i-th diffraction pattern. The CCF $C^{i,i}(q, \Delta)$ can be analyzed using a Fourier series decomposition [8–10],

$$C^{i,i}(q, \Delta) = C_0^{i,i}(q) + 2\sum_{n=1}^{\infty} C_n^{i,i}(q)\cos(n\Delta), \tag{3}$$

where $C_n^{i,i}(q)$ are the Fourier components of the CCF, particularly $C_0^{i,i}(q) \equiv \langle I^i(q, \varphi)\rangle_{\varphi}^2$, and $C_n^{i,i}(q) = |I_n^i(q)|^2$ for $n \neq 0$.

For a disordered system, the Fourier components $C_n^{i,i}(q)$ fluctuate from position to position on the sample [8–12], therefore averaging of the CCF (or its Fourier spectra) over a large number M of diffraction patterns is applied to get a statistically averaged result,

$$\langle C^{i,i}(q,\Delta)\rangle_M = 1/M \sum_{i=1}^{M} C^{i,i}(q,\Delta). \tag{4}$$

Another two-point CCF $C^{i,j}(q,\Delta)$ can be defined as an angular intensity correlation function between two different diffraction patterns i and j,

$$C^{i,j}(q,\Delta) = \left\langle I^i(q,\varphi) I^j(q,\varphi+\Delta)\right\rangle_\varphi, \quad i \neq j, \tag{5}$$

where $I^i(q,\varphi)$ and $I^j(q,\varphi)$ indicate the intensities measured on the i-th and j-th diffraction patterns, respectively. Similar to $C^{i,i}(q,\Delta)$, this CCF can be expressed in terms of the Fourier components $C_n^{i,j}(q)$ and averaged over a set of diffraction patterns, $\langle C^{i,j}(q,\Delta)\rangle_M = 1/M \sum_{i,j}^{M} C^{i,j}(q,\Delta)$, where the summation is done over M random pairs of diffraction patterns with $i \neq j$. Following derivations presented in [8–12] it can be shown that $C_n^{i,j}(q) = I_n^{i*}(q) \cdot I_n^j(q)$, where the Fourier components of intensity $I_n^i(q)$ and $I_n^j(q)$ are defined for the i-th and j-th diffraction patterns. In the absence of correlation between two different diffraction patterns i and j, an average of the product reduces to the product of averages and we obtain for the Fourier components $\langle C_n^{i,j}(q)\rangle_M$ averaged over M random pairs of diffraction patterns $\langle C_n^{i,j}(q)\rangle_M = \langle I_n^{i*}(q) \cdot I_n^j(q)\rangle_M = \langle I_n^{i*}(q)\rangle_M \cdot \langle I_n^j(q)\rangle_M$. It can be shown [8–12] that for random orientational distribution of particles each of the contributions, $\langle I_n^{i*}(q)\rangle_M$ and $\langle I_n^j(q)\rangle_M$, is proportional to a so-called random phasor sum at $n \neq 0$ and approaches zero for sufficiently large M. At these conditions $\langle C_n^{i,j}(q)\rangle_M \to 0$ for $n \neq 0$. At the same time nonvanishing values of $\langle C_n^{i,j}(q)\rangle_M$ indicate the presence of correlations between diffraction patterns. Such correlations may appear, for example, due to different sources of the background scattering that could not be easily determined. This property of $C^{i,j}(q,\Delta)$ is essential to detect contribution to the scattered intensities that can not be attributed to the sample. In this work we analyze the difference Fourier spectra $\langle \widetilde{C}_n(q)\rangle_M = \langle C_n^{i,i}(q)\rangle_M - \langle C_n^{i,j}(q)\rangle_M$ that contain only structural information intrinsic to the sample.

3. Experiment

The coherent x-ray scattering experiment was performed at the nanoprobe endstation GINI-X [16] installed at the coherence beamline P10 of the PETRA III facility at DESY in Hamburg. The scattering geometry of the experiment is shown in Fig. 1(a). The incident photon energy was chosen to be 13 keV and a 2D detector was positioned in transmission geometry at 195 mm distance from the sample and protected by a beamstop of 34 mm in diameter. The scattering data were recorded on a hybrid-pixel detector Pilatus 1M from Dectris with 981×1043 pixels and a pixel size of $172 \times 172\ \mu m^2$.

The sample was mounted on a goniometer, and a film was aligned with its surface perpendicular to the direction of the incident beam. The beam with a flux of about $5 \cdot 10^{10}$ photons/sec was focused on the sample with Kirkpatrick-Baez (KB) mirrors to a spot of about $200 \times 300\ nm^2$ (FWHM). The sample was scanned in the plane perpendicular to the incident beam direction with a step size larger than the probe size. A cryogenic cooling of the film with liquid nitrogen was used during measurements to reduce sample damage. The exposure time was chosen to be 0.3 sec per image to perform measurement in non-destructive regime.

The two samples studied in the experiment were spin-cast from a blend of P3HT with gold nanoparticles using a similar fabrication protocol. P3HT material (molecular weight 44900 g/mol; PDI 1.22) has been synthesized at the University of Wuppertal, Germany. Gold

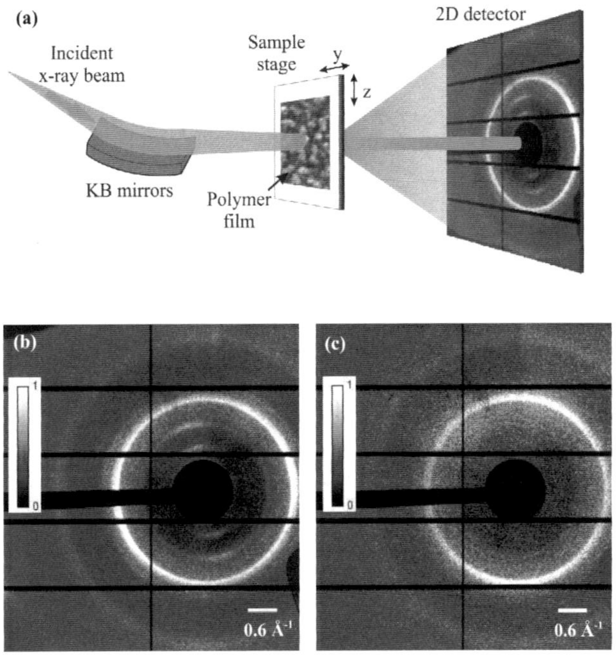

Figure 1. (a) Geometry of the scattering experiment showing the focusing KB mirrors, sample, and 2D detector. (b),(c) Typical diffraction patterns measured for two different P3HT polymer films with Au nanoparticles.

nanoparticles stabilized with fluorene derivatives (AuNPs-SFL) have been prepared in analogy to literature reports [17–19]. A 5 mg/ml blend solution (P3HT:AuNPs-SFL = 10:1 by weight) in chloroform was spin cast on rectangular shaped grids with a 15 nm thick Si_3N_4 membranes (Dune Sciences, Inc.).

4. Results

We measured an x-ray dataset for each sample, consisting of 800 and 1600 diffraction patterns respectively. Measured data were corrected for background scattering and polarization of incident x-rays. Typical snapshot diffraction patterns corresponding to the two samples are shown in Figs. 1(b) and 1(c). Radial intensities $\langle\langle I(q,\varphi)\rangle_\varphi\rangle_M$ averaged over the full dataset for each sample in the range $0.6 \leq q \leq 2.0$ $\mathring{A}^{-1}$ are presented in Figs. 2(a) and 2(c). As one can see, both intensity curves show similar features. The strongest scattered intensity visible on both diffraction patterns is a (020) ring at $q = 1.65$ $\mathring{A}^{-1}$, which is related to the π-π stacking of conjugated network[1].

The major differences between the diffraction patterns shown in Figs. 1(b) and 1(c) can be attributed to a different orientation of (020) planes and alkyl side chains in the corresponding samples. In Fig. 1(b), relatively strong (200) and (300) peaks at $q = 0.8$ $\mathring{A}^{-1}$ and $q = 1.2$ $\mathring{A}^{-1}$ are observed due to mixed orientation of alkyl side chains, that are tilted with respect to the sample surface [the (100) reflection was not accessible in our experiment because it was covered by the beamstop]. While the intensity distribution around the (020) peak has a ring-like structure due to randomly distributed domains, the (200) and (300) peaks are strongly anisotropic. This is

[1] There is also a ring, that partially visible on the detector at higher scattering angles ($q = 2.7$ $\mathring{A}^{-1}$), which corresponds to scattering from gold nanoparticles. Due to a small amount of AuNPs in the polymer host matrix this scattering is rather weak.

reflected in the two-fold rotational symmetry of the intensity distribution at lower **q** [Fig. 1(b)]. In contrast to this, the $(h00)$ peaks are hardly visible in Fig. 1(c). This indicates that alkyl chains are mostly aligned perpendicular to the sample surface (edge-on orientation), while π-π stacking is parallel to the surface producing the (020) ring. Note, the ensemble-averaged radial intensity [Fig. 2(c)] shows the $(h00)$ peaks comparable to the case illustrated in Fig. 2(a). At this level, similar averaged features can be accessed by GIXD technique.

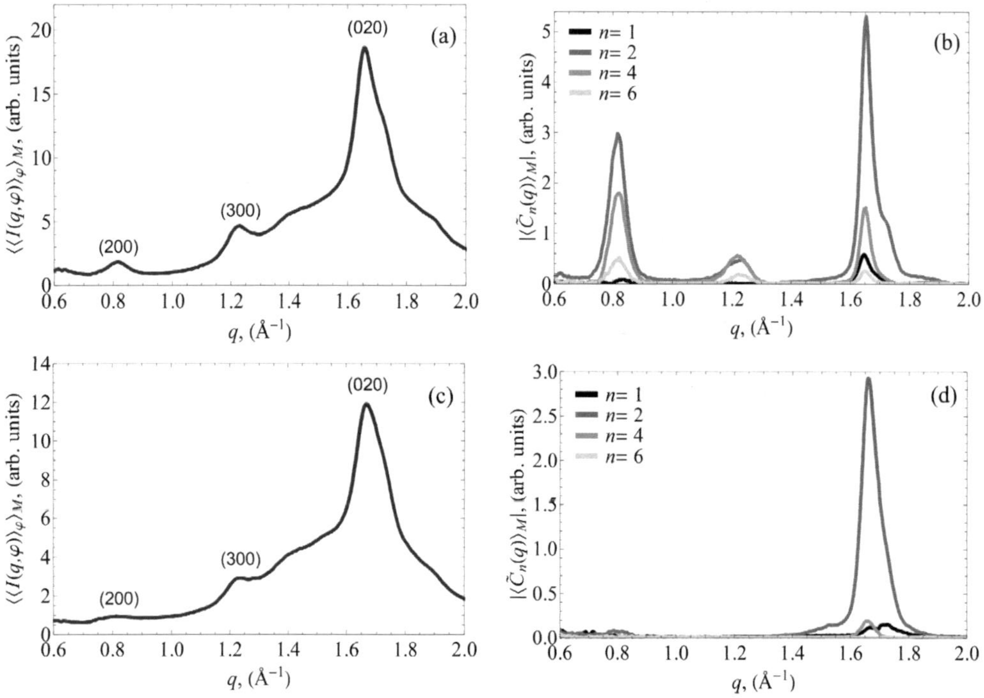

Figure 2. (a),(c) Ensemble averaged radial intensity $\langle\langle I(q,\varphi)\rangle_\varphi\rangle_M$ and (b),(d) modulus of the difference spectra $|\langle\widetilde{C}_n(q)\rangle_M|$ calculated for two different samples.

Next, we performed the cross-correlation analysis on both datasets. The resulting difference Fourier spectra $\langle\widetilde{C}_n(q)\rangle_M$ are presented in Figs. 2(b) and 2(d). Although these Fourier spectra were averaged over many diffraction patterns, they are defined by average *nanoscale* structural properties of the sample due to the nano-sized x-ray probe applied in the experiment. For the first sample the Fourier spectrum $\langle\widetilde{C}_n(q)\rangle_M$ [Fig. 2(b)] has three local maxima for the Fourier components of the orders $n = 2, 4$ and 6. These maxima are located at the same q values as (200), (300) and (020) peaks on the average intensity curve $\langle\langle I(q,\varphi)\rangle_\varphi\rangle_M$ [Fig. 2(a)]. The presence of higher harmonics of $n = 2$ (up to $n = 6$) indicates a high degree of orientational order (similar to bond-orientational order parameters observed in hexatic phase of a liquid crystal [15]). For the second sample the Fourier spectrum $\langle\widetilde{C}_n(q)\rangle_M$ has a strong maximum only at the position corresponding to (020) ring. In this case, the strongest contribution is given by the Fourier component[2] $n = 2$, whereas $n = 4$ is significantly weaker, and $n = 6$ is about noise level.

These results indicate a coupling of the orientational anisotropy of different interplanar distances associated with π-π stacking and alkyl side chains. Interestingly, for the mixed orientation of domains the same Fourier components (with $n = 2, 4$ and 6) contribute to Fourier

[2] The Fourier component of the order $n = 1$ also gives a non-zero contribution to the spectra in Figs. 2(b) and 2(d). The origin of this contribution will be clarified in the future work.

spectra of the CCF at $\mathbf{q}$ corresponding to all ($h00$) and (020) peaks [see Fig. 2(b)]. The Fourier components of higher orders $n = 2$ and 4 are related to a higher symmetry of the conjugated system induced by AuNPs. One can also see a slight difference of the shape for the (020) radial intensity peak as well as for the most intense $n = 2$ Fourier component for two samples [see Figs. 2]. A possible reason for the broadening of the (020) peak in the case of mixed orientation may be due to a contribution of the higher order (400) peak and P3HT backbone related planes.

5. Conclusions

We applied the XCCA for the analysis of x-ray scattering data from two P3HT films with addition of gold nanoparticles. Despite similar fabrication protocol for both films, probing with nanobeam on the local scale reveals structural differences between two systems. While the radial intensity curves look similar for two samples, the Fourier spectra of the intensity cross-correlation functions clearly indicate different structural properties of the systems. These results strongly support a pronounced variation of local order upon addition of Au nanoparticles. It confirms inhomogeneous structure of the samples, with preferential edge-on orientation of domains in one sample and mixed orientation in another sample. For the sample with dominating edge-on orientation of domains the main contribution is given by $n = 2, 4$ components, and for the film with mixed orientation an additional contribution to the Fourier spectra with $n = 6$ is clearly observable, indicating overall higher ordering of organic system confined by AuNPs. The obtained results demonstrate that XCCA provides complementary information about the structure of disordered samples that is missing in the standard GIXD analysis.

Acknowledgments

We are thankful to the group of T. Salditt, especially to M. Osterhoff and S. Kalbfleisch, for providing support of Göttingen instrument for nano-imaging with x-rays (GINI-X), and also A. Zozulya for the help during the experiment. We acknowledge fruitful discussions and support of this project by E. Weckert. We thank J. A. Sellberg and T. J. Lane for discussions concerning correlation functions, and S. Funari for a careful reading of the paper. We gratefully acknowledge Maria Vittoria Russo, Rome and Ullrich Pietsch, Siegen for helpful discussions. Part of this work was supported by BMBF Proposal 05K10CHG "Coherent Diffraction Imaging and Scattering of Ultrashort Coherent Pulses with Matter" in the framework of the German-Russian collaboration "Development and Use of Accelerator-Based Photon Sources" and the Virtual Institute VH-VI-403 of the Helmholtz Association. This work was also partially supported by the Department of Chemistry, Sapienza University of Rome through the Supporting Research Initiative 2013 and BMBF (project Nr 05K10PSC).

References

[1] Salleo A, Kline R J, Delongchamp D M and Chabinyc M L 2010 *Adv. Mater.* **22** 3812–3838
[2] Prosa T J, Winokur M J, Moulton J, Smith P and Heeger A J 1992 *Macromolecules* **25** 4364–4372
[3] Rahimi K, Botiz I, Stingelin N, Kayunkid N, Sommer M, Koch F P, Nguyen H, Coulembier O, Dubois P, Brinkmann M and Reiter G 2012 *Angew. Chem. Int. Ed. Engl.* **51** 11131–11135
[4] Kim D H, Jang Y, Park Y D and Cho K 2006 *Macromolecules* **39** 5843–5847
[5] Sirringhaus H and et al 1999 *Nature* **401** 685–688
[6] Shabi T S, Mikayelyan E, Grigorian S, Pietsch U, Koenen N, Scherf U, Kayunkid N and Brinkmann M 2012 *Macromolecules* **45** 5575–5585
[7] Wochner P, Gutt C, Autenrieth T, Demmer T, Bugaev V, Diaz-Ortiz A, Duri A, Zontone F, Grübel G and Dosch H 2009 *Proc. Nat. Acad. Sci.* **106** 11511–11514
[8] Altarelli M, Kurta R P and Vartanyants I A 2010 *Phys. Rev. B* **82** 104207
[9] Altarelli M, Kurta R P and Vartanyants I A 2012 *Phys. Rev. B* **86** 179904(E)
[10] Kurta R P, Altarelli M, Weckert E and Vartanyants I A 2012 *Phys. Rev. B* **85** 184204
[11] Kurta R P, Dronyak R, Altarelli M, Weckert E and Vartanyants I A 2013 *New J. Phys.* **15** 013059
[12] Kurta R P, Altarelli M and Vartanyants I A 2013 *Adv. Cond. Matt. Phys.* **2013** 959835

Journal of Physics: Conference Series **499** (2014) 012021 doi:10.1088/1742-6596/499/1/012021

[13] Gutt C, Grodd L, Mikayelyan E, Pietsch U, Kline R J and Grigorian S (in preparation)

[14] Parashchuk O, Grigorian S, Levin E, Bruevich V, Bukunova K, Golovnin I, Dittrich T, Dembo K, Volkov V and Paraschuk D 2013 *J. Phys. Chem. Lett* **4** 1298

[15] Kurta R P, Ostrovskii B I, Singer A, Gorobtsov O Y, Shabalin A, Dzhigaev D, Yefanov O M, Zozulya A V, Sprung M and Vartanyants I A 2013 *Phys. Rev. E* **88** 044501

[16] Kalbfleisch S, Neubauer H, Kruger S P, Bartels M, Osterhoff M, Mai D D, Giewekemeyer K, Hartmann B, Sprung M and Salditt T 2011 *AIP Conf. Proc.* **1365** 96–99

[17] Fratoddi I, Venditti I, Battocchio C, Polzonetti G, Bondino F, Malvestuto M, Piscopiello E, Tapfer L and Russo M V 2011 *J. Phys. Chem. C* **115** 1519815204

[18] Matassa R, Fratoddi I, Rossi M, Battocchio C, Caminiti R and Russo M V 2012 *J. Phys. Chem. C* **116** 15795–15800

[19] Quintiliani M, Bassetti M, Pasquini C, Battocchio C, Fratoddi I, Rossi M, Mura F, Matassa R and Russo M V (in preparation)

AUTHOR INDEX

AUTHOR INDEX

Institute of Physics
Dirac House, Temple Back
Bristol BS1 6BE UK

CURRAN ASSOCIATES INC
proceedings.com